The Stars in their Courses is a collection of this famous author of science and science fiction. They are concerned with such diverse topics as astronomy, overpopulation and the possibility of time-travel. Addressed to the general reader as well as to the scientist, they are written in Dr. Asimov's familiar, informative and entertaining style. *The Stars in their Courses* is an invaluable addition to the bookshelves of anyone who wishes to keep abreast of scientific thought.

Also by Isaac Asimov in Panther Books

Foundation
Foundation and Empire
Second Foundation

Earth is Room Enough
The Stars Like Dust
The Martian Way
The Currents of Space
The End of Eternity
The Naked Sun
The Caves of Steel
Asimov's Mysteries
Nightfall One
Nightfall Two

I, Robot
The Rest of the Robots

The Early Asimov: Volume I
The Early Asimov: Volume II
The Early Asimov: Volume III

Nebula Award Stories 8 (ed)

Isaac Asimov

The Stars in their Courses

Panther

Granada Publishing Limited
Published in 1975 by Panther Books Ltd
Frogmore, St Albans, Herts AL2 2NF

First published in Great Britain by White Lion
Publishers 1974
Copyright © 1969, 1970 by Mercury Press Inc
Copyright © 1971 by Isaac Asimov
All essays in this volume appeared in
The Magazine of Fantasy and Science Fiction
between May 1969 and September 1970
Made and printed in Great Britain by
Richard Clay (The Chaucer Press) Ltd
Bungay, Suffolk
Set in Monotype Plantin

Dedicated to:
My old friend, Lester del Rey,
and to the memory of
Evelyn del Rey

CONTENTS

INTRODUCTION

I saw the play *Hair* recently, largely because everyone told me I just *had* to see it. The music, the joy, the youthful verve, the delight — Adjectives were thrown at me as though they were darts and I was a dartboard.

So I went —

The first thing that happened was that the actors and actresses spread themselves all over the stage, the theater fixtures and the audience, and started mooing.

'Aries,' bleated someone in a husky whisper.

'Leo,' groaned someone else.

'Gemini,' 'Libra,' 'Sagittarius' came from here and there and everywhere. They were intoning the signs of the zodiac and a cold dread crept over me, for this could only mean that I was about to be immersed in a sticky sea of irrational folly.

The dread was justified! There came a chant about Jupiter being in the house of Mars (or something) and everyone started a dithyramb about the 'Age of Aquarius.'

After that, it was all downhill. Partly through natural square-hood and partly through a dogged devotion to rationality, I kept trying to make sense out of what I could hear through my shattered eardrums. I didn't manage.

When people find out now that I saw *Hair*, a look of holy ecstasy crosses their face (for they will lose their ticket to the land of With-It if they don't register approval) and they say, 'How did you like it? Wasn't it great?'

I nod vigorously and say, 'It was very, very loud!'

That sounds like approval so that I avoid any further discussion – and I do that without telling a lie.

But what is this 'Age of Aquarius'? I don't know how many young people snap their fingers, sway, and look orgastic as they moan its words, but do they know what it is they're singing? I haven't found any who did. The question even seemed to annoy some. It was as though noisemaking is fun but knowing is a drag.

So I'll tell you what the Age of Aquarius is.

The vernal equinox is the moment when the Sun crosses the Celestial Equator on its apparent motion northward, and it usually falls on March 20 on our calendar. If the Earth's axis of rotation were absolutely motionless relative to the stars, the Sun would be at the same point in the sky at every vernal equinox.

However, because of the pull of the Moon's gravity on the equatorial bulge of the Earth, the Earth's axis of rotation shifts in such a way that the North Celestial Pole (and the South Celestial Pole, too) marks out a circle in the sky; a circle with a radius of 23.5° of arc. It takes about 25,780 years for the poles to complete the circle.

This means that every year, because of this shift in direction of the Earth's axis, the Sun's position at vernal equinox is displaced westward just a little bit and the time of the vernal equinox precedes the time of arrival it would otherwise have. The phenomenon is therefore called the 'precession of the equinoxes.'

All right so far?

The vernal equinox finds the Sun crossing the Celestial Equator at a point in one of the twelve signs of the zodiac. Gradually, though, its location moves westward from one sign into the next, then into the next, and so on. In 25,780 years it will have passed through all twelve signs of the zodiac. It spends 2150 years (on the average) in each sign. Of course, there is no sharp moment of passage from one sign to the next because there are no clear boundaries between the signs, these being purely human conventions rather than astronomic fact.

At the time astrology was invented by the ancient Sumerians the vernal equinox found the Sun in the constellation which we now call Taurus (the Bull). By the time the Greeks took up astrology in a big way, the Sun had moved into Aries (the Ram) at the time of the vernal equinox.

The Greeks naturally began the list of signs of the zodiac with Aries, giving it domination over the period from March 20 to April 19. Astrologers today still follow the Greek fashion, applying Aries to that period of the year and the subsequent signs to the corresponding following monthly periods.

Of course, about two thousand years ago, the vernal equinox edged into Pisces (the Fish), but astrologers pay no attention to the fact that all their signs are now given to the wrong months. It is the advantage of mysticism that, having no logical content, it can't be damaged in any way by any further increase in nonsense, however great.

And in the not-too-distant future, the vernal equinox will edge its way into Aquarius (the Water Carrier). It is in that sense that we are entering or are about to enter the Age of Aquarius.

But what effect will the alteration have on Earthly affairs? What will the stars bring about?

Nothing, of course. See the first two chapters of this book.

At least, it is not the location of the Sun relative to the stars that will do anything for us or to us. But something, nevertheless, may happen (and very likely will) in the near future that will reduce all previous crises to nothingness in comparison. See the last two chapters of this book.

And if so, then *Hair* deeply depresses me. This is no time for young people to turn from science to mysticism; from reason to emotion; from the honest sweat of concern to the loose verbalism of 'love.' To find refuge from the miserable reality of today in the euphoria produced by drugs or mysticism (not very far apart in their effects) is to surrender – to lie down and wait for death.

To be sure, science and reason have not saved us yet. Indeed, in their tragic service to the criminal forces of unreason (see Chapter 15) they have intensified our troubles. Yet there remains no substitute for them.

Science may be, and has been, misused, but the proper cure for that is not to replace science by non-science and sense by non-sense – but to replace science-misused by science well-used.

It is to science (and therefore to life) that I am devoted and it is the delight of the pursuit of science that I proselytize – in this book among others.

A
Astronomy

1

The Stars in their Courses

One of the pitfalls to communication lies in that little phrase 'It's obvious!' What is obvious to A, alas, is by no means obvious to B and is downright ridiculous to C.

For instance, just a week ago, a storekeeper was writing out a receipt for me. He asked my name and I gave it and, as I do automatically, began to spell it, slowly and clearly. (I am absurdly sensitive about having my name misspelled.)

He got the I-S-A-A-C right but by the time I had gotten as far as A-S-I- in the spelling of the last name he had raced ahead and finished it as M-O-F.

'No, no, no,' I said, pettishly, 'it ends with a V, a V.'

He changed it to A-S-I-M-O-V, looked at it a moment and then said, 'I see. You spell it the way the author does.'

Well, obviously I do, but it came as a big surprise to him that I chose to do so.

That's a small item; here's a bigger one. If anyone asks me what I think of astrology, I say something like, 'It's stuff and nonsense, sheerest bilge, absolute tripe. Obviously!'

Except that there's nothing obvious about it to most people.

Astrology is more popular today than ever before in history and more people than ever make a good living out of it. I have read that there are five thousand astrologers in the United States and over ten million true believers.

There was a time when I could have shrugged that off and said something like: 'Oh, well, finding that one American out of twenty is gullible and unsophisticated is no great shock.'

But the greatest popularity explosion in astrology right now is among the college students who, one might suppose, are the well-read, the intelligent, the sophisticated, the hope of the future.

The question arises then: If the collegiates are taking up astrology, how can it be 'obviously' absolute tripe?

It can be tripe with no trouble at all. Consider —

(1) It is fashionable right now, especially among college students, to oppose the Establishment; that is, to take up a position directly antagonistic to the one accepted by the leaders of some particular segment of society. Some young men do it out of considered thought and honest emotion and I sympathize. (I'm rather anti-establishment in some ways myself, for all that I'm a little over thirty and am approaching late youth.)

Let's face it, though. Many college students oppose the Establishment just because it is fashionable in their set to do so and for no other reason. The opposition is quite blind as far as they are concerned and they could easily be manipulated into crew cuts, for instance, if President Nixon would only make the supreme sacrifice and grow long hair.

Well, there is such a thing as a scientific Establishment, too. There is an accepted canon of scientific thought which says (among other things) that the characteristic quality of astrological lore is very like the excrement of the male bovine, and that is enough reason for many an ardent youngster to become an enthusiastic astrologian.

(2) We live in troubled times. To be sure, all times are troubled (as some bland pundit is certain to say at this point) but none has ever been quite as troubled as ours is. When before have we had the inestimable pleasure of knowing that one hasty move can blow up the world in a half-hour display of thermonuclear temper tantrum? When before have we had the exciting choice of being brought to chaos and destruction by either overpopulation or overpollution within half a century without anyone being sure which will win the race?

Yet in this tottering society of ours, science has no pat answers. It has only a program of procedure, a system for asking questions and testing the answers for validity – with the very good chance that said answers will prove invalid. Opposed to this are various systems of mysticism which give answers loudly, clearly and confidently. Wrong answers, to be sure, but what's the difference?

13

The sad thing for us rationalists is that the vast majority of the human race would rather be told that 'Two and two is five and make no mistake about it,' than 'I think it is possible that two and two may be four.'

(3) College students are no more a homogeneous group than is any other large classification of humanity. Not all of them are interested in science; not all of them are truly bright. Many of them are just bright enough to discover that what counts in this phony world is merely the ability to *sound* bright – an ability which has carried many men to high political office.

It is easier, they soon learn, to sound bright in some subjects than others. It is, for instance, just about impossible to sound bright in mathematics or the physical sciences without actually being bright. The facts, observations and theories are too well established. There is a firm consensus and you have to know a great deal about that consensus before you can sound bright, and for that you have to *be* bright.

The consensus is shakier in the social sciences; still shakier in the humanities; and in matters such as mystical Eastern cults (just to take an example) there is no consensus at all.

Someone who spouts nonsense in chemistry will be caught at once by any high-school student who knows something about chemistry. Someone who spouts nonsensical literary criticism, however, can be spotted only with difficulty. Indeed, what are the criteria for nonsense in literary criticism? Do you know? Does anyone?

As for mysticism, hah! Your bluff in this field cannot possibly be called. Make up a chant such as: Toilet Tissue, Toilet Tissue, Toilet, Toilet, Tissue, Tissue — Tell everybody that this chant repeated 666 times (the number of the beast) will induce inner serenity and cosmic consciousness and you will be believed. Why not? It sounds no worse than anything else in mysticism and you will become a highly respected swami.

To put it as briefly as possible: Many college students are taking up astrology in a big way because (1) it is the in thing to do, (2) it gives them a delicious, if false, sense of security, and (3) it gives them a passport to phony intellectualism.

14

And none of that is at all inconsistent with astrology being tripe.

The funny part is that astrology started off as the best science man could find.

In man's cultural dawn, when the Universe seemed a whimsical place and the gods were constantly hitting one over the head without good reason, there had to be some system for finding out what those troublesome divinities wanted. Desperately, priests sought answers by watching the flight of birds, the shape of the livers of sacrificed animals, the fall of dice and so on.

These events were essentially random in nature, but early man did not recognize the principle of randomness. (Many of our contemporaries don't either.) All events were either man controlled or god controlled, and if a particular event was not man controlled it therefore had to be god controlled. So people study tea leaves and head bumps and palm creases even today.

A great advance was made by certain priests (probably in Sumeria, a land which later became Babylonia, then Chaldea, Mesopotamia and finally Iraq). If I may be allowed to reconstruct what their reasoning may have been, here it is.

The gods, they may have argued, could scarcely be so inefficient and so wasteful in time and effort as to make special messages for each occasion. How ungodly to take the trouble to create particular livers or to send a particular bird flying in a particular direction, or to go to the trouble of thundering in this quarter of the sky or that, every time there was something to say.

A truly great god would scorn such trivia. He would instead create some natural phenomenon that was continuous and yet complex – a kind of moving finger that would steadily write the history of the world in all its facets and would act as adviser for man. Instead of man depending on the uncertainties of special revelation, he would merely have to work out the laws that governed the continuous but orderly complexity of the natural phenomenon.*

* If the early astrologers argued in this fashion, and very possibly they did, they were imbued with the spirit of science and I honor

15

The one natural phenomenon that was absolutely steady and inexorable, that could apparently be set into motion once and for all, was the movements of the heavenly bodies.

The Sun rose and set day after day and, at its daily peak, shifted north and south in a slower rhythm. The Moon rose and set day by day and changed phases in a slower rhythm. The mathematical rules describing these changes were not simple, but were not so complex that they could not be worked out.

Furthermore, these shifts clearly affected the Earth. The Sun caused the alternation of day and night by its rising and setting, and the slower alternation of seasons by its movement north and south. The Moon's rising and setting (along with its phases which were easily seen to be related) gave successions of lighter nights and dimmer ones. (Its phases were also related to the tides – a fact of extraordinary importance – but for various reasons this wasn't firmly noted till the end of the seventeenth century.)

Obviously, if the shifting Sun and Moon could affect conditions on Earth, then an 'astrologic code' must exist. If you can predict the change in shifts in the heavens, you ought to be able to predict the change in conditions on the Earth.

Of course, it is rather trivial to predict that tomorrow morning the Sun would rise and the Earth would light up, or that the Moon was waning and the nights would grow dark, or even that the noonday Sun was shifting southward and that cold weather was therefore on its way. All that was simple enough for ordinary men to handle but it lacked detail. Would there be enough rainfall? Would the crops prosper? Would there be war or pestilence? Would the queen have a baby son?

For that the sky had to be studied in greater detail.

We will never know what early observer or observers began making systematic observations of the position of the Moon and the Sun against the starry background. The thousands of stars maintained their relative position night after night, year after year, generation after generation (so that they were called

---

them. No scholar can be maligned for being wrong in the light of the knowledge of a later period. If he strives for knowledge in the terms of his own time he is a member of the brotherhood of science.

'fixed stars'), but the Sun and Moon shifted position with respect to them. Eventually, the Greeks called them 'planets' ('wanderers') because they wandered among the stars.

Both Moon and Sun took a certain fixed path among the stars, the two paths being fairly close together. They traveled at different rates, however. While the Sun made one complete circuit around the heavens, the Moon managed to make twelve. (Actually twelve and a fraction, but why complicate matters?)

It was useful then to mark off the path by means of easily detected signposts – a development probably started by the Sumerians but brought to perfection by the Greeks.

Suppose you take a strip of stars in a circle around the heavens; the particular strip of stars that contains the paths of the Sun and the Moon; and divide it into twelve equal parts or 'signs.' Start the Sun and the Moon in the same place in one of those signs. By the time the Moon has gone around once and returned to the sign, the Sun has moved one twelfth of the way around and shifted into the next sign. Another circuit of the Moon would find the Sun shifting into the next sign after that.

As an aid to the memory, draw patterns among the stars in each sign, preferably patterns that resemble familiar animals, and you have twelve constellations making up the 'zodiac' ('circle of animals').

Once you start to study the zodiac carefully, you are bound to discover five bright stars that are *not* fixed but that wander around the zodiac as the Sun and Moon do. These are five more planets, in other words, and we know them today as Mercury, Venus, Mars, Jupiter and Saturn.

These new planets add immeasurably to the complications of the heavens and, therefore, to the potential of the 'astrologic code.' Some of them move quite slowly. Saturn, for instance, makes only one complete circle of the sky while the Moon is making 360 of them. What's more, while the Sun and Moon always move from west to east against the background of the stars, the other planets sometimes shift direction and briefly move east to west in what is called 'retrograde motion.' Saturn does it no less than twenty-nine times during the course of a single revolution.

I stress the point that the early astrologers were not fakers. If they had been charlatans, it would have been much easier to stick to watching birds and livers.

Establishing the background of astrology meant watching the skies night after night, making painstakingly accurate observations and, in short, working one's head off. And what they discovered was factual and valuable. Their observations represented the beginnings of real astronomy and have remained a completely valid description of the machinery of the solar system (relative to a stationary Earth) to this very day.

Where the astrologers went wrong was not in their description of the heavens but in their working out of the 'astrologic code.' And even here they must have done their best to be rational.

Where could you find a clue to the code? Suppose there was an event in the heavens that was extremely rare. Would that not mean that any equally rare event that followed on Earth would be related to it? And could not one learn something from the relationship?

For instance, suppose there was an eclipse. Suppose it was one of those rare occasions when the Moon was slowly blotted out of the heavens; or one of those even rarer occasions when the Sun was. Would not that be followed by some equally notable event on Earth?

The question almost answered itself, for eclipses struck absolute panic in the hearts of all who watched, and understandably so. It is routine to laugh at that panic, but don't. Suppose you knew very well that your life depended on the Sun and suppose you watched the Sun slowly fading before an encroaching blackness for reasons you could not explain. Would you not feel the Sun was dying? And that all life would die with it?

(Consider that in our own 'sophisticated' time it is only necessary for some solemn idiot to proclaim that California will fall into the Pacific Ocean on 3 P.M. of next Thursday to cause a quick exodus of uncounted thousands from that state.)

Well, then, if an eclipse is so rare and frightening a phenomenon, it is very easy and almost inevitable to argue that its consequence must be an equally rare and frightening event on Earth. In short, an eclipse must portend disaster.*

But never mind theory. Does disaster follow an eclipse in actual fact?

Sure, it does. In any year with an eclipse there is some catastrophe somewhere and this is easy to see since in *every* year, eclipse or not, there is some catastrophe somewhere.

Astrologers seize upon the catastrophes that follow eclipses. Unscientific? Certainly. But very human. (In this very enlightened year in which we live, try arguing with someone who firmly believes that lighting three on a match is unlucky. Tell him that misfortunes happen even when only lighting two on a match and see how far you'll get.)

As it happens, it could not have taken very long for the early astrologers to learn the cause of eclipses. They would note that the Moon was eclipsed whenever it was on the side of the Earth directly opposite the Sun, and therefore in the Earth's shadow. They would note that the Sun was eclipsed whenever it and the Moon were precisely in the same spot in the sky and we were therefore in the Moon's shadow.

By carefully calculating the motions of the Moon and the Sun, it was possible to predict lunar eclipses in advance without too much trouble. (Some think that the ancient Britons used Stonehenge for that purpose in 1500 B.C.) Solar eclipses were harder to calculate but eventually they, too, could be handled.

It is easy to see that astrologers would be tempted to keep their methods secret. The common folk wouldn't be able to follow the calculations anyway and would be annoyed if they were asked to. Besides, the astrologers probably found their social standing greatly enhanced and they managed easily to keep it so without letting on that anyone could do it if he would but take the trouble to master the mathematics involved.

Of course, it had its risks. A Chinese legend reports that in very ancient times an eclipse came to the capital without warning because the royal astronomers Hsi and Ho, preoccupied with a drinking bout, somehow neglected to let it be known that

---

* The very word 'disaster' is from a Greek term meaning, essentially, 'evil stars.'

it was going to happen. After the emperor had gotten over his imperial fright at the unexpected event, the suddenly sobered astronomers were led off to execution and all agreed that it was richly deserved.

An eclipse could have more beneficent results, too. Farther west, in ancient times, the Sun's disc was encroached upon by darkness, little by little, over a field of battle in Asia Minor. The armies of Lydia on the west and Media on the east stopped fighting and peered at the vanishing Sun. The few minutes of eclipse-night came and when they had passed the opposing generals could do only one thing. They signed a treaty of peace and went home. Lydia and Media never fought again, for they knew the anger of the gods when they saw it.*

As it happens, modern astronomers can calculate the exact date of the eclipse of the Sun that took place in Asia Minor at about that time. It was on May 28, 585 B.C., so that the Lydian–Median battle is the earliest event in all history, other than the mere fact of an eclipse, that can be pinned down to an exact day.

The Greek philosopher Thales was supposed to have predicted the eclipse, though not to the exact day – merely that one would take place that year. He is supposed to have traveled in Babylonia in his youth and he probably learned the prediction trick from astronomers there.

There was another astronomical event that broke the quiet routine of the heavens, and that was the coming of a comet.

It created even worse terror than that caused by an eclipse, taken over all, for several reasons.

Whereas an eclipse came and went in a relatively short period of time, a comet would remain in the sky for weeks and months. Whereas an eclipse involved perfectly regular shapes (arcs of circles), comets had weird and ominous forms – a fuzzy head with a long tail that might look like a sword suspended over the

* This perpetual peace is not as impressive as it might be for both nations disappeared from the map about thirty years later when Cyrus of Persia conquered them. No doubt if they had endured longer they would eventually have forgotten the lesson of the eclipse and gone back to war.

Earth, or the disordered hair of a shrieking woman. (The very word 'comet' is from a Greek word for 'hair.')

Finally, whereas an eclipse could be predicted even in ancient times, the coming of a comet could not be. A system for predicting the arrival of *some* comets wasn't worked out till the eighteenth century.

Comets were even surer indications of catastrophe then than eclipses were, and were indeed followed by catastrophes for the same reason.

Thus in 1066, the comet we now call Halley's Comet appeared in the sky just as William of Normandy was making ready to invade England. It predicted catastrophe and that is exactly what came, for the Saxons lost the Battle of Hastings and passed under the permanent rule of the Normans. The Saxons couldn't have asked for a better catastrophe than that.

On the other hand, if the Saxons had won and had hurled William's expeditionary force into the Channel, that would have been catastrophe enough for the Normans.

Whichever side lost, the comet was sure to win.

With eclipses and comets serving so excellently to predict events on Earth, the principle of the 'astrologic code' seemed well established and the technique, too, for it seemed to work on the principle of similarity. A disappearing Sun bespoke disappearing prosperity; a comet with a tail like a sword bespoke war, and so on.

With the Greeks, democracy invaded astrology. In the East, the philosophy of the oriental monarchies, where only the king counted, kept astrology the handmaiden of high political affairs. Among the individual-centered Greeks, the personal horoscope came into use.

One could imagine them arguing that since the Sun was the brightest of the planets (using the word in the ancient sense) it had the most to do with the individual. In which sign was the Sun at the moment of that individual's birth? If it was in the constellation of Libra (the Scales), ought he not to be of even and judicious temperament; if it was in Leo (the Lion), ought he not to prove a brave warrior?

If you stop to think that the ancients thought the heavenly bodies were small objects quite close to the Earth, and the constellations somehow really represented the things they seemed to represent, it all makes a weird kind of sense.

Even so, there were two important groups in the palmy days of the Greeks who opposed astrology.

The Greek philosophical school of Epicureanism opposed it because their view of the Universe was essentially an atheist one. They felt the heavenly bodies moved purposelessly and that no gods existed to weave meaning into their motions.

The other group was that of the Jews, who were unusual among the people of the time, for being cantankerously monotheistic. They were not scientifically minded and they used no rational argument to oppose astrology. (They would have been unspeakably horrified at the Epicurean reasoning.) It was just that those who supported astrology were pagans and considered the planets to be gods and this was anathema, on principle, to the Jews.

Yet even the Jews were not wholly uninfluenced by astrology. The older writings that appear in the Bible were carefully edited in Greek times by pious rabbis intent on wiping out unedifying traces of a polytheistic past – but the erasures weren't perfect.

Thus, on the fourth day of Creation, the Bible states: 'And God said, let there be lights in the firmament of the heaven to divide the day from the night; and let them be for signs, and for seasons, and for days and years.' (Genesis 1:14.) That little word 'signs' is an astrologic hangover.

A clearer one is to be found in the Song of Deborah, one of the oldest passages in the Bible, an ancient poem too well known to endure much tampering. After the defeat of Sisera, Deborah sang: 'They fought from heaven; the stars in their courses fought against Sisera.' (Judges 5:20.)

Neither the Epicureans nor the Jews prevailed, however. Astrology continued and was exceedingly popular in the seventeenth century when modern astronomy came gloriously into its own. In fact, some of the very founders of modern astronomy – Johannes Kepler, for instance – were astrologers, too.

But by the end of the seventeenth century, with a true picture of a heliocentric solar system established, astrology finally became a pseudo science. It passes human understanding to suppose that the vast Universe we now recognize is arranged only as a key for our own insignificant dust speck. That so many men and women believe it is, just the same, is a remarkable tribute to the manner in which human folly can triumph over all.

Still, science has its prestige even among its enemies. There are those devotees of astrology who know just enough about real astronomy to seek some legitimate scientific rationale for the pseudo science.

And such is the ingenuity of man, particularly when it is misapplied, that such a rationale (extraordinarily weak but a rationale, nevertheless) can actually be found. I'll talk about that in the next chapter.

2

The Lop-sided Sun

Last fall, a large-circulation magazine wanted to get an article on eclipses that they could run in conjunction with a solar eclipse that was to take place the following spring over the United States.

It occurred to the magazine to get me to do the job and you can bet I was willing to do so. Writing about eclipses was apple-pie-and-mother for me, and I had never appeared in the pages of this particular magazine and I wanted to do so.

But the fact that I had never done anything for the magazine was naturally a source of insecurity for the editors. They wanted to talk to me and asked me to come visit their offices.

I did so, and listened to them explain very carefully what they wanted. I nodded and said I understood and that I would try to strike the very note they were asking for.

But then one of the editors thought a bit and said, 'Could you describe a total eclipse for us? How would it look? What would you see? What is it like?'

'All right,' I said, and calling on the skills that years and years of writing had made second nature, I described a total eclipse in the most moving imaginable terms. By the time I was through I had them (and myself) all but dissolved in tears.

'Good,' they said. 'Write that article for us, just as you told it.'

I did, and they liked it (and said so), paid for it, published it and all was absolutely well.

In fact, during the whole transaction, there was only one moment during which a bit of nervousness made itself felt. While I was describing the total eclipse in poignant detail, I was uneasily aware that all could be ruined by a single, simple question.

What if someone in my audience had said, 'But, Dr. Asimov, have you ever seen a total eclipse?'

For of course I hadn't.

But then, as I told you, I called upon my writing skills and I was a fiction writer to begin with.

Naturally, I must not allow a narrow escape like that to shake my nerve so, like the thrown rider climbing back on the horse, I will now turn, unrepentant, to the Sun again.

To anyone who is so foolish as to try to look at the Sun in its glory, it must seem to be an eternally featureless circle of brilliant white light. Indeed, there have been theologians who maintained that if it weren't exactly that, it would represent a flaw in the perfection of God's handiwork, and who therefore resisted any suggestion that such flaws might exist.

It was one of the unsettling aspects of Galileo's astronomical discoveries that, in 1610, he reported the existence of spots on the face of the Sun. Once that troubling fact was reported, and despite considerable clerical shock, others saw them at once. Indeed some spots are huge enough to be made out with the unaided eye. When the Sun is very close to the horizon on a particularly clear day, and is ruddily dim enough to be looked at without harm, large sunspots can sometimes be made out.

Sunspots are easy to see because they are regions that are cooler than the surrounding solar surface and that therefore seem dark in comparison. Black-on-white is impossible to miss.

But what about the reverse situation? What if there are local regions of the Sun's surface that are *hotter* than surrounding areas? If so, they would be unusually bright, but whiter-on-white is a lot harder to see than black-on-white and, in point of fact, no one saw hot regions on the Sun for two and a half centuries after the cool regions were detected.

The honor of the later discovery belongs to Richard Christopher Carrington, an English astronomer who kept painstaking track of sunspots over prolonged periods of time, working out the exact time of rotation of the Sun at different latitudes. (A gaseous body does not rotate all in one piece as a solid body perforce must.)

In 1859, Carrington noted a short-lived, brilliant flare-up on the face of the Sun. It was as though a tiny star had made itself visible on that face for some five minutes. Carrington reported it and suggested a cause. At this time, astronomers were considering the possibility that the Sun might use as the source of its radiation the kinetic energy of impacting meteors, and Carrington felt he had been lucky enough to see the impact of a particularly large meteor.

It was a very interesting guess, but wrong.

If whiter-on-white is hard to see in general, this may not be so at all wavelengths. That is, the general increase in brightness in a particular hot spot may, for some reason, be greater at some wavelengths than others. If the Sun were viewed by the light of the wavelength particularly affected, a flare-up that would be difficult or even impossible to see over the entire spectrum might suddenly become unmistakably conspicuous.

In 1889, the American astronomer George Ellery Hale invented the 'spectroheliograph,' a device whereby the Sun could be photographed by light coming through a spectroscope so arranged that all light, except over a small stretch of wavelengths, is excluded. In that way, the Sun could have its picture taken by hydrogen light or by calcium light. Seen by calcium light it was easy to find out, for instance, that there were calcium-rich regions here and there on the Sun, standing out like clouds in a summer sky on Earth.

Using the spectroheliograph, one tends to get static pictures of the Sun and misses short-lived events; that is, it is difficult to tell whether a particular spot on the photograph has just come into being or is soon to pass – unless you take a number of closely spaced stills.

In 1926, Hale (still alive, still active, still ingenious) devised a modification of the instrument that enabled one to watch events by the light of a spectral line over a period of time. This 'spectrohelioscope' made it a lot easier to detect swift changes.

Before the 1920s were out, then, it became apparent that, by hydrogen light, there were flare-ups rather commonly associated with sunspots. There were what seemed to be explosions, sudden flashes of hot hydrogen, that might be at full heat for five to ten minutes and be utterly gone after half an hour to an hour.

These were 'solar flares' and, looking backward, it was understood that what Carrington had seen seventy years before had been an unusually bright one.

When a flare is on the side of the Sun facing us, there's not much to be seen but a brightening and spreading patch. Occasionally, though, one catches a flare coming into being at the edge of the Sun. Then one can see, in profile, a huge surge of brilliant gas climbing at a rate of six hundred miles a second or so, and reaching a height of some five thousand miles above the Sun's surface.

Small flares are quite common and in places where there is a large complex of sunspots, as many as a hundred a day can be detected, especially when the spots are growing. Very large flares of the kind that approach visibility in white light (like Carrington saw) are rare, however; and only a few occur each year.

The spectra associated with these flares indicate temperatures of up to 20,000° C. as compared with 6000° C. for the undisturbed surface of the Sun and 4000° C. for the dark center of sunspots.

Solar flares are important in connection with a more general activity of the Sun's surface. Energy is somehow transferred from the Sun's glowing surface to the thin solar atmosphere, or 'corona.' That energy must be distributed among the atoms of the corona which are far fewer in number than are those of the surface. This means that the energy per atom is far higher in the corona than on the Sun's surface and 'energy per atom' is what we mean by 'temperature.'

It is not surprising then, that where the surface temperature of the Sun is 6000° C., the coronal temperatures can be as high as 2,000,000° C. The intensity and wavelength of radiation from any body depends upon its temperature and the corona delivers more radiation (per unit mass) than the Sun's surface does. It is only because the corona has so small a mass that it seems so faint. What's more, coronal radiation is far more energetic than surface radiation is, and it is from the corona that solar X rays arise.

Nor is electromagnetic radiation all that flows out of the Sun. The turbulent solar atmosphere sends matter streaming upward and small quantities of it inevitably manage to escape even the Sun's tremendous gravity. There is a constant drizzle of particles moving outward and, apparently, lost to the Sun forever.

In absolute terms the mass of particles lost in this fashion is enormous by Earthly standards, for it comes to a million tons per second. By Solar standards it is nothing, for if this loss were to continue indefinitely at its present rate, it would take six hundred trillion years for the Sun to lose 1 per cent of its mass.

These particles, spreading outward from the Sun in all directions, make up the so-called 'solar wind.'

The solar wind extends to the Earth and beyond, of course, but the Earth's small globe intercepts only a tiny part of all the particles cast out by the Sun. Of the million tons of particles lost by the Sun each second, about three fourths of a pound strike the Earth. This is not much in terms of mass, but it means that every second something like a hundred trillion solar particles reach the vicinity of the Earth.

If the Earth were without atmosphere or magnetic field, those particles that reach the vicinity of the Earth would go on to strike the surface of the planet. They strike the surface of the Moon, for instance, and the samples of rock brought back by the astronauts contain quantities of helium that can have originated only in the solar wind.

The particles in the solar wind are naturally representative of the material in the Sun. The Sun is very largely hydrogen, with most of what is left being helium. At the temperature of the corona through which the solar wind passes, atoms of hydrogen and helium are broken down to a mixture of atomic nuclei and electrons. The hydrogen nucleus is a proton and the helium nucleus, an alpha particle.

The protons are much more massive than the electrons and much more numerous than the still more massive alpha particles, and if both mass and number are taken into account, it is clear that the major components of the solar wind are its protons. Any increase in the density of the wind due to something happening back on the Sun may be called a 'proton event.'

Since the Earth has a magnetic field, the electrically charged particles of the solar wind (one positive charge for protons, two for alpha particles and one negative charge for electrons) are deflected along the magnetic lines of force. That means they move in a tight spiral from one magnetic pole to the other, back and forth over and over. It is these moving particles, held in place by the magnetic lines of force that make up what used to be called the Van Allen belts but are now more often called the 'magnetosphere.'

The magnetosphere dips closest toward the Earth's surface at the magnetic poles and it is there that the charged particles most easily leak out of the magnetosphere and into the Earth's upper atmosphere. The interaction of the charged particles and the atoms of the upper atmosphere produces the shifting curtains and streamers of the aurora.

Well, then, what happens when a flare lights up a portion of the solar surface? There is a localized rise in temperature and a localized increase in turbulence that result in the sending of a blaze of energy and a rash of particles into the corona immediately above the flare. The coronal temperature rises and there is an increase in its production of ultraviolet radiation and X rays at the affected spot. The additional rush of particles also produces a kind of gust in the solar wind so that the solar flare could, in effect, result in a proton event.

The intensification of the solar wind above a particularly large flare can be so great that the speeding protons become energetic enough to count as mild cosmic rays.

If the solar flare shoots up into the Sun's atmosphere in the direction, more or less, of Earth, there is a burst of energetic radiation toward us that reaches our planet in minutes, and there is also a gust in the solar wind that reaches us in a couple of days. When that gust of solar wind reaches the Earth's magnetic field, there is a sudden brightening and extending of the aurora.

The radiation from the flare and the subsequent flood of charged particles upsets the situation in Earth's upper atmosphere. It may produce wild static in electronic equipment or it may wipe out (temporarily) some of the charged layers in the upper atmosphere, causing radio waves to pass upward into space instead of being reflected downward toward the ground.

Radio transmission can then fade out altogether, and radar may grow useless.

These manifestations are usually termed 'magnetic storms' because one of the symptoms is a wild irregularity of the needle of the magnetic compass in response to all the jolts being undergone in the region of the magnetic poles.

Variations in the magnetic compass and intensification of the aurorae are interesting but not, in today's society, very significant. The possible disruption of radio transmission is another matter. It can seriously annoy our electronically oriented population and industry. At particularly pathological moments (say, during wars or threats of war) the possibility that radar may go awry, that radio-controlled missiles may wander off course, and that communications of all sorts may be distorted or destroyed can be a serious source of worry.

Then, too, astronauts in space or on the Moon's surface may be caught in the aftermath of a flare, be subjected to an intense gust of the solar wind and suffer radiation sickness.

With this in mind, it would naturally become a matter of great interest to be able to predict the coming of flares long enough in advance to keep men out of space, to protect men on the Moon, and to set up alternate methods of communication in a war zone.

It would help if we knew what caused flares in the first place, but we don't. Since flares are characteristically found in connection with sunspots, we might suppose that if we knew what caused sunspots, we might deduce what caused flares – but we don't know what causes sunspots, either.

But suppose we reason like this—

Sunspots represent an asymmetry on the Sun. A sunspot forms at some particular place on the Sun's surface and not on others. Why should that happen? Why shouldn't all parts of the Sun's surface be alike? After all, the Sun is a nearly perfect sphere and, as nearly as we can theorize, it is radially symmetrical. That is, working from the center outward, properties change equally no matter what direction we choose.

On Earth, we have weather. We have storms developing in one place and calm in another; zephyrs here and tornadoes there; drought yonder and floods elsewhere. But all this is the result of a tremendous asymmetry – the fact that one side of the Earth faces the Sun's heat and the other does not, at any given time, and that even on the side facing the Sun there are variations in the length of time of exposure and the direction from which the Sun's rays are received.

On the Sun, however, there is no such overwhelming asymmetry in evidence. Why, then, does it have 'weather' in the form of sunspots?

To be sure, the Sun rotates, so there is a difference with respect to the centrifugal effect as related to latitude. There is no centrifugal effect at the pole and maximum effect at the equator, with intermediate values in between. The rotation may also set up asymmetries in the deeper layers of solar material, too.

It would not be surprising then if the sunspot appearance was somehow connected with latitude, and that turns out to be correct. Sunspots tend to appear only between 5° and 30° north and south latitude.

Within that latitude range there is a complex regularity. The sunspots increase in numbers to a maximum, then decrease to a minimum, then increase to a maximum again, with a period of eleven years. Immediately after a minimum, spots appear at about 30° north and south latitude. Then, as they increase in numbers from year to year, they tend to shift toward the equator. At periods of maximum, they are at an average latitude of 15° and continue to shift toward the equator as they decrease in number again. As the cycle dies within 5° of the equator, a new set begins to appear at 30°.

Nobody knows why the cycle works in that way, but can it be that it is the result of some asymmetry more complicated than that introduced by the Sun's rotation? If so, where might that asymmetry come from? One possibility is that it is imposed from the outside and the finger of suspicion points to the planets.

But how can the planets affect the Sun? Surely only by way of their gravitational fields. Those fields might raise tides on the Sun and make it lopsided.

The Moon, for instance, raises tides on the Earth because the side of the Earth near the Moon receives a stronger lunar pull than the side on the opposite side of the Moon. It is this difference in pull that produces the tidal effect.

The size of the tidal effect depends upon three things. First, upon the mass of the tide-producing body, of course, since the greater the mass, the greater the gravitational pull. Second, upon the diameter of the body experiencing the tides, since the greater the diameter, the greater the difference in gravitational pull experienced on opposite sides. Third, upon the distance of the body experiencing the tides from the tide-producing body. The greater the distance, the weaker the gravitational pull of the latter on the former, and the smaller the difference in pull on the two sides of the former.

Taking all these factors into account, we can set up the following table:

System	Relative tidal effect
Moon on Earth	7000
Sun on Earth	3200
Earth on Sun	1

The tidal effect of the Earth on the Sun is only a ten-thousandth of that of the Sun and Moon combined on Earth, but perhaps this effect, though tiny, is not entirely insignificant.

What about the tidal effects of other planets? Taking the Earth's tidal effect on the Sun as unity, it isn't difficult to work out the relative tidal effects of the other planets on the Sun. Here are the results:

Planet	Relative tidal effect on Sun
Mercury	0.7
Venus	1.9
Earth	1.0
Mars	0.03
Jupiter	2.3
Saturn	0.11
Uranus	0.0021
Neptune	0.0006
Pluto	0.000002

The four greatest tidal effects, then, are those of Mercury, Venus, Earth and Jupiter. All the remaining planets put together produce a tidal effect about a fifth that of Mercury, the least of the big four. We might therefore give the name 'tidal planets' to Mercury, Venus, Earth and Jupiter.

As these planets circle the Sun, each produces a pair of tiny bulges on the Sun (one bulge on the side of the Sun toward itself, another on the opposite side). These bulges may be tiny indeed, measured perhaps only in centimeters, but even that much rise and fall of the Sun's vast surface could be significant.

With four separate bulges on each side of the Sun, continually changing positions with respect to each other as the planets move, there might just conceivably be some crucial moment when, as a result of the combination of bulges, a particular section of the Sun's surface rises or falls with unusual speed. Can it be, then, that it is at that point and at that time that the sequence of changes that produces a sunspot is set in motion? Perhaps, too, there is some long-range pattern in the shifting bulges that accounts for the sunspot cycle and the regular shift in latitude.

Of course, it is hard to try to correlate the appearance of a particular sunspot with a particular combination of tidal bulges. The appearance is too slow. But what about flares? Flares come and go quickly and sizable flares are not very common. Can large flares be correlated with not-very-common planetary positions?

Perhaps!

Dr. J. B. Blizard, a physicist at the University of Denver, has studied the connection of proton events with planetary conjunctions; that is, with situations in which two or more of the tidal planets take up such positions that a line through them points toward the Sun. The tidal effect would then add up; there would be an unusually high bulge in the Sun's surface and perhaps if that passes near a region of sunspot turbulence, it would set off a flare.

In any case, over the period between 1956 and 1961, Blizard noted a sufficient number of positive correlations; that is, of flares coming near the time of conjunction, to make matters look interesting. He calculated that there was only one chance in two thousand of the correlations being just coincidence. What's more, he began a series of predictions of the occurrence of flares in the future, supposing them to come at the time of later conjunctions (which are easy to calculate) and achieved 60 per cent accuracy.

And here, then, is where we can return to the subject of astrology, which I discussed in Chapter 1. I said in that chapter that astrologers would naturally seize on any scientific rationale, however faint, for justifying their folly, and I suspect that Blizard's work will be right up their alley.

For years now, astronomers and non-astronomers have wondered about the effect of the sunspot cycle on Earth. Was there more solar radiation at sunspot minimum? Was there sufficiently more, perhaps, to produce droughts, cut down the food supply, affect prices, start inflations or depressions, make wars more likely, and so on?

When flares were discovered, they seemed even more likely than sunspots themselves to affect Earth. After all, they bathed the atmosphere in charged particles and that might affect rainfall which in turn might affect crops which in turn might affect— Besides, who knows what more subtle changes might rise and fall with the quantity of charged particles striking Earth?

Now comes along Dr. Blizard who makes it seem that the rise and fall in the solar wind may depend upon planetary position and 'planetary position' is a magic phrase to astrologers. I can look forward to the following chain of reasoning.

1 – The planetary positions control the occurrence of solar flares.

2 – Through the solar flares, the planetary positions control the nature of the solar wind.

3 – Through the solar wind, the planetary positions control and effect certain subtle changes on Earth.

4 – Because of all this, the planetary positions at the moment of a child's birth may affect all the course of its future life.

I am sure that astrologers will not hesitate to use this line of reasoning to justify the specificity of the following items taken from the astrological column of a recent edition of my newspaper:

'Aries. Give sympathy, not money, to friend in financial dilemma. This loan won't be repaid.'

'Scorpio. Don't trust investment tip; there are factors at work your adviser doesn't know about.'

To be sure, there may be some among my valued Gentle

Readers who will be swayed by this and who will wonder whether there isn't *something* in this chain of reasoning starting from planetary position and ending in Aries giving sympathy and Scorpio turning on the distrust.

If so, repress the thought. I am quite certain that with very little ingenuity I could invent a chain of reasoning that is just as valid and plausible, connecting the pattern of burping of a herd of hippopotami amid the reeds of the river Nile with the rise and fall of the steel output in the mills of Gary, Indiana.

3

The Lunar Honor-roll

In my youth, my father discovered that science fiction was my favorite reading matter. Memory stirred within him and he said to me.

'Science fiction! Going to the Moon! Aha! Tell me, did you maybe ever read books by Zhoolvehrn?'

I stared at him blankly. '*Who?*'

'Zhoolvehrn,' he repeated.

I was rather chagrined. I flattered myself that I knew the important writers of the world, together with the important *and* unimportant science-fiction writers, and it annoyed me to be found wanting.

'What did he write?' I asked.

'Science fiction. Going to the Moon, and so on. Oh, and he wrote a book about a man who went around the world in eighty days.'

Light broke with blinding brilliance. I knew the author well, but my father had never heard the name pronounced in anything but the French fashion. I said (and in the excitement my stately Brooklyn accent became a trifle more prominent than usual), 'Oh, sure. The author you mean is Joolz Voin.'

And my father said, '*Who?*'

Anyway, however we might be divided by a common language, it turned out that my father and I both enjoyed science fiction. So it was a particular delight to me that when Neil Armstrong set foot upon the Moon he had done it not only within my own lifetime but even within my father's lifetime.

As an amazing way of dramatizing the speed with which technology is driving onward, consider that when my father was born (on December 21, 1896) no man had ever in all the history of the Earth lifted himself from the ground in powered flight. There were balloons and gliders, but these were passive devices for floating on air and/or making use of the wind.

It was not till July 2, 1900, when my father was three and one-half years old, that the first directed and completely controlled flight took place. (No, the date is not a mistake and I'm not talking about the Wright Brothers.)

The inventor in question was Ferdinand, Count von Zeppelin. He conceived the notion of confining a balloon within a cigar-shaped structure of aluminum, making it both sturdier and aerodynamically more efficient. Beneath it he suspended a gondola bearing an internal-combustion engine which served to drive a propeller and to pull gondola and balloon through the air *even against the wind*.

Zeppelin had invented the zeppelin (what else) or the 'dirigible balloon,' meaning 'balloon capable of being steered in flight.' Inevitably, the latter phrase was shortened to 'dirigible.'

On December 17, 1903, just a few days before my father's seventh birthday, the Wright Brothers flew their airplane, and that was the first controlled flight of a *heavier-than-air vehicle*.

On March 16, 1926, when my father was twenty-nine years old, Robert Hutchings Goddard sent up the first liquid-fueled rocket. The rocket traveled only 184 feet but it was a portent of things to come. By 1944, when my father was forty-seven years old, much larger liquid-fueled rockets, launched by Wernher von Braun, were bombarding London.

On October 4, 1957, a rocket-propelled vehicle was, for the first time, placed in orbit about the Earth, and my father was sixty years old then. On April 12, 1961, when my father was sixty-four, the first man-carrying vehicle was placed in orbit around the Earth.

And finally, on July 20, 1969, when my father was seventy-two and one-half, human footsteps appeared on the soil of the Moon.

Mankind had gone from a state of imprisonment on Earth's surface all the way to the Moon, and had done it all in the course of one not unusually long lifetime.

We must expect that this rapid sweep of technology will continue. There will be additional trips to the Moon; longer stays on its surface; a greater variety of experiments conducted there; and eventually the beginnings of a permanent base on the Moon which may and should develop into a colony.

In the course of all that, the names of various features on the Moon will become familiar to all who read the newspapers and watch television. This will be good, for a number of romantic names will come into their own as well as a number of great and good men of the past. This will also be bad, for there will be ample opportunity for television announcers and others to mangle those names beyond recognition.*

But before the names become utterly banal through overuse, let's go through some of them.

The notion that it would be necessary to name features upon the Moon was not conceivable before June 1609, when Galileo looked at the Moon with his telescope. It was only then that astronomers realized that the Moon did not have a shiny, flat, polished and (barring certain smudges) featureless surface. (This featurelessness had till then been assumed, in accordance with Aristotle's dictum of the perfection of the heavens.) Instead, the Moon had mountains and valleys and, in general, a surface at least as rough and various as that of the Earth.

Galileo drew the first map of the Moon, showing a few craters and one or two of the large dark areas. Succeeding astronomers, with better instruments, saw details more clearly and the maps began to improve. It also became more tempting to give names to the various features.

The first to draw maps of the Moon that were so good that we can actually detect the features of the lunar (loo′ner) surface as we recognize them today was the German astronomer Johannes Hevelius.

In 1647, he published a magnificent volume called *Selenographia*, which was an atlas of the Moon's surface. He titled the features systematically, but avoided using personal names

* I am irritated for instance, by the tendency of announcers to treat the word 'lunar' as though it is an exotic recent invention. It is an old word which is pronounced *lyoo′ner* or, by the less precise, *loo′ner*. The announcers, however, say *loo′nahr′*, giving the syllables equal weight and exaggerating the second vowel as in 'radar.' I wonder how they would pronounce 'popular,' 'vulgar' and other words of the sort that end in 'ar' and are pronounced as though they end in 'er.' For that matter, do they speak of the *soh′lahr′* system?

for fear of the envy and backbiting that might result. Instead he followed the new view that the Moon was, after all, but a smaller Earth, and transferred the names of geography bodily into selenography. The various lunar mountain ranges were given the Earthly names of 'Alps,' 'Apennine Mountains,' 'Carpathian Mountains,' 'Caucasian Mountains,' and 'Taurus Mountains.'

These names have remained to this day but, of course, this can give rise to confusion. It is customary even now, and will undoubtedly become obligatory later, to speak of the 'Lunar Alps,' the 'Lunar Apennines' and so on.

Hevelius also named the large, dark areas 'seas.' By that time, it was already quite apparent that the likelihood of the Moon possessing surface air or water was low, but Hevelius was intent on using as many Earthly names as possible.

Happily, he did not name the seas literally for those on Earth, but indulged in fanciful flights.

He used Latin in his naming of course, and called a sea a 'mare' (pronounced rather like Mary in English), and the plural for this is 'maria' (with the accent on the first syllable). It is just as well, for the seas are not seas in any Earthly sense, and it is much less confusing to speak of them, in general, as maria, even though the names of specific examples may still use 'sea' for reasons of convenience and (yes) poetry.

Thus, the area on which the astronauts of Luna 11 first landed was on the edge of Mare Tranquillitatis, which can be translated into English, most picturesquely, as the 'Sea of Tranquillity.' Considering the Moon's unchanging landscape (barring the occasional strike of a meteorite, the occasional split of a rock due to temperature change, the occasional outwelling of gas, powder or lava from some rifts in the ground) this is a delightfully apt name. Almost immediately adjacent is the Mare Serenitatis or 'Sea of Serenity.'

Other names are far less apt, though it is unlikely that any will (or should) be changed even though most refer to the water that is conspicuously lacking on the Moon. Thus, without trying to be exhaustive, we have:

Mare Imbrium ('Sea of Showers')
Mare Nectaris ('Sea of Nectar')

Mare Humorum ('Sea of Moisture')
Mare Spumans ('Sea of Foam')
Mare Vaporum ('Sea of Steam')
Mare Undarum ('Sea of Waves')

Most inappropriate of all is the Mare Foecunditatis ('Sea of Fertility').

One particularly large mare is the Oceanus Procellarum ('Ocean of Storms'). The small dark areas get correspondingly smaller titles. There is the Lacus Somniorum ('Lake of Dreams'). There are also the Sinus Iridum ('Bay of Rainbows') and Sinus Roris ('Bay of Dew') separated by a ridge of high ground.

A few features are accurately named. For instance, a flat area in the very middle of the visible face of the Moon is Sinus Medii ('Central Bay').

Where the real trouble will come, will be in the names of the craters. They are untranslatable and a number are a trifle unpronounceable.

The fault lies with an Italian astronomer named Giovanni Battista Riccioli who, in 1651, published a book called *New Almagest*, in which he included his own maps of the lunar surface – maps which were not as good as Hevelius', by the way.

Riccioli departed from Hevelius' system of avoiding personalities and began the practice of naming the craters for dead astronomers (and other prominent people). To understand what he did, it is important to see what his astronomical beliefs were.

The key point is this: He rejected the views of Copernicus, who placed the Sun at the center of the planetary system. Riccioli was a conservative who clung as closely as possible to the time-honored and hoary astronomical views of the Greeks. He considered the Earth as the center of the universe and believed the heavenly bodies to be moving in perfect circles. He knew of Kepler's theory that the planets (and the Earth, too) traveled in ellipses about the Sun but dismissed that without deigning any argument. To those who pointed out that the Sun-centered Copernican system was preferable because it was far simpler than the older Earth-centered Ptolemaic system, Riccioli countered that the more complicated the system, the better its testimony to the greatness and glory of God.

Back in 1577, the Danish astronomer Tycho Brahe had proposed a compromise. He suggested that all the planets (Earth excluded) moved about the Sun in circular paths as Copernicus suggested, but that the Sun *with* its circling planets moved about the Earth. This system captured many of the virtues of the Copernican system without abandoning the basic Greek assumption of the Earth as center of the Universe.

As a compromise it was virtually stillborn, but a few conservatives clung to it for dear life as the only alternative to a complete abandonment of Earth-centered doctrine (an abandonment they feared would have important theological overtones). Riccioli was one of those who clung to the Tychonic modification of the Earth-centered view.

Well, Riccioli's naming of the craters reflected his estimate of the relative worth of astronomers.

Thus, there are three particularly prominent craters on the Moon; three that were perhaps the most recently formed of the large ones. Each is surrounded by a system of rays; straight lines of light material that go streaking out from the crater in all directions and which seem to be dust sprayed out from the crater in its formation. All large craters probably had rays associated with them but in time those rays were obliterated by later strikes. In the case of these three I'm talking about there seem to have been no later strikes of great account.

By far the most prominent of the three is near the lunar south pole (near the top of the globe in most photographs, which usually have south on top and north below).

When the sunlight strikes straight down on this crater, it becomes exceedingly bright. It and its ray system stand out from their surroundings to the point where the Moon looks like a navel orange with the crater itself serving as the navel. Or, if one is exceedingly naive, it looks as though the Moon has a literal 'north pole' with visible meridians. The crater is the most magnificent single object on the Moon, both on the visible side and the hidden side.

What would Riccioli call that crater, do you suppose? One

would scarcely need to be uncertain, in view of his predilections. He named it Tycho (tie'koh, in English).

The other two craters, also magnificent but far less so than Tycho, he named Copernicus and Kepler. Still, let's not complain for if he had less scientific integrity, he might have omitted those Sun-centered gentlemen altogether.

Again, near the center of the Moon's visible disc, is a grouping of four large craters located at the corners of an imaginary diamond-shaped figure.

The largest of these, Riccioli called Ptolemaeus (tol''uh-mee'us) after Claudius Ptolemaeus (better known in English as Ptolemy) who in the second century summarized the work of the Greek astronomers in a book that was one of the few that survived from ancient times. It was often referred to by the admiring Greeks as 'Megiste' ('Greatest') and the Arabs affixed their own definite article and made it 'Almagest.' Notice that Riccioli called his own book *New Almagest*.

Immediately to the northeast of Ptolemaeus, on the opposite angle of the diamond, is a somewhat smaller crater which Riccioli named Hipparchus (hip-pahr'kus). Hipparchus was the greatest of all the Greek astronomers and though his own writings are lost they formed the chief basis on which Ptolemy worked out his system. Hipparchus was the first to work out the Earth-centered view of the Universe in full and satisfactory mathematical detail, but it was Ptolemy, the summarizer, rather than Hipparchus, the creator, who gave his name to the since-called 'Ptolemaic system.' And it is Ptolemy who gets the bigger crater, too. There are vast injustices in scientific history, as elsewhere.

On the other two angles of the diamond are a pair of craters named for medieval supporters of the Ptolemaic view. One is Albategnius (al''buh-teg'nee-us), which is the Latinized version of the name of a tenth-century Arabic astronomer, al-Battani, the greatest of all medieval astronomers.

The other, just south of Ptolemaeus and, indeed, encroaching upon it, is Alphonsus (al-fon'sus), which is named for a Castilian monarch, Alfonso X, usually called 'Alfonso the Wise.' He was an unsuccessful king from the politico-military standpoint, but he was noted for his scholarship, for his encouragement of learning, for the schools he founded and the law codes he organized.

Under his patronage the first history of Spain was written and Jews of Toledo prepared a new and particularly good translation of the Old Testament. He himself wrote poetry, as well as alchemical commentaries.

His place on the Moon arises from the fact that he encouraged the preparation of revised planetary tables, tables that could be used to predict the location of the planets at any given time, past or future. (Ideally, at least, but in practice they were increasingly inaccurate for larger and larger intervals of time in either direction.) They were published in 1252, on the day of his accession to the throne. These 'Alfonsine Tables' proved the best the Middle Ages had to offer and were not replaced by better ones for over three centuries.

Riccioli might have had a few doubts, perhaps, as to the propriety of giving Alfonso so prominent a crater. After all, the royal scholar did cast doubt on the Ptolemaic system. During the tedious preparation of the tables on the basis of the complicated mathematics made necessary by insisting on Earth-as-center, Alfonso is supposed to have said, in exasperation, that had God asked his advice during the days of the creation, he would have strongly recommended a simpler design for the Universe.

The ancient astronomer who is most likely to impress moderns is the fourth-century B.C. Aristarchus (ar″is-tahr′kus), if for no other reason than the modernity of his views.

He was the first to make an accurate measurement of the distance of the Moon and he attempted to make one of the Sun as well. His method of measurement was perfectly correct in theory but he was hampered by the lack of proper instruments that would enable him to reach sufficient precision, and his estimate of the solar distance therefore fell far short of the truth.

Aristarchus was the first to advance the suggestion that the planets, *including* the Earth, all revolved about the Sun. For his pains, he was laughed at heartily, and at least one philosopher (Cleanthes, the Stoic) demanded he be tried for impiety.

Aristarchus' works have not survived since few scribes would get the necessary fee required to copy over those crackpot theories and the only way in which we know about those theories is that other Greek philosophers refer to them sneeringly.*

The views of Aristarchus have survived through these references throughout the Middle Ages. Copernicus seems to have known of them since he mentioned them in a passage of the manuscript he was writing; a mention he later cautiously crossed out.

One wonders, then, why we speak of the Copernican system rather than the Aristarchean. In this case, though, it is not a matter of injustice; Copernicus deserves the credit. Although Aristarchus had the right idea, he did not work out the mathematics of planetary motion on the basis of a Sun-centered system. One of the reasons that the Greeks turned to Hipparchus and his Earth-centered system was that Hipparchus supplied the necessary mathematics for his view.

When Copernicus came along he, *for the first time*, supplied astronomers with the necessary Sun-centered mathematics and that is why he deserves the credit.

Riccioli had the grace to name a crater for Aristarchus, but his prejudices show. Where Hipparchus and Ptolemy got large, centrally located craters, Aristarchus got a small one far to the north-west.

The largest crater clearly visible on the side of the Moon facing us is Clavius (klay'vee-us). This honor is granted a competent German astronomer, much honored in his own time, but virtually unknown now. His chief virtue in Riccioli's eyes was, of course, that he rejected the Copernican system.

Riccioli did not use the names of astronomers only in marking his craters. He also employed the names of politicians and other notables toward whom he felt sympathetic and whom he thought ought to be honored.

* Of course, I like to make disparaging remarks about crackpot notions (or what I consider crackpot notions) myself as in Chapters 1 and 4. I know that there is a danger that someday someone will say something like, 'The only way we know of so-and-so's important and world-shaking theories is from Asimov's sneering references —' but I'll take that chance.

Since Riccioli's time, additional craters have been named for luminaries who have lived after him, and the concentration has been heavily on scientists, preferably astronomers.

The map of the Moon has thus become a resounding listing (an honest honor-roll) of astronomical accomplishment. There is a heavy representation, thanks to Riccioli and others, of ancient philosophers. In addition to those already mentioned, here are a few who are to be found among the craters: Anaxagoras (an″ak-sag′oh-ras), Anaximander (uh-nak′sih-man′der), Anaximenes (an″ak-sim′ih-neez), Archimedes (ahr″kih-mee′-deez), Aristoteles (ar-is-toh′tih-leez, better known to us as Aristotle), Eratosthenes (er″uh-tos′thih-neez), Euclides (yoo-kligh′deez, better known to us as Euclid), Eudoxus (eu-dok′-sus), Philolaus (fil″oh-lay′us), Posidonius (pos″ih-doh′nee-us), Pythagoras (pih-thag′oh-ras), and Thales (thay-leez).

What a field of mispronunciation.

There are some Arabic astronomers preserved in the crater honor-roll that are wonders, too. How about Arzachel, for instance. He was an eleventh-century Moslem astronomer in Spain and his proper Arabic name was Ibn al-Zarqala. I'm not at all sure how one pronounces Arzachel; my guess is ahr-zak′el, but I could easily be wrong.

There are relatively modern crater names that can give trouble, too. In the eighteenth century, there was a French scholar named Jean Sylvain Bailly, who wrote important histories of astronomy. He also participated in the French Revolution and was mayor of Paris in 1789. French politicians in those days made being guillotined a kind of hobby and Bailly was no exception. He got it in the neck in 1793.

In time, a crater was named for him; a large one, even larger than Clavius, but so far toward the edge it couldn't be made out clearly till the days of the rocket probes.

The name Bailly is pronounced, French fashion, as bah-yee′, but we can very safely bet that no American on the Moon will call the crater anything but bay′lee.

Among the modern astronomers listed on the lunar honor-roll are Bessel, Bond, Cassini, Flammarion, Flamsteed, Herschel, Huggins, Lassell, Messier, Newton and Pickering. Of famous men who are *not* primarily astronomers there are, for example:

Cuvier, Guericke, Gutenberg, Herodotus and Julius Caesar. There's also a crater named for Riccioli himself and one which, in ecumenical fashion, is named Rabbi Levi.

The Soviets continued Riccioli's system in naming the craters on the other side of the Moon, and introduced one very important innovation. They named a crater for a science fiction writer – Jules Verne (Zhoolvehrn, Joolz Voin, take your pick).

I don't want to suggest that this be carried on to ridiculous extremes,* but I think it's fair to hope that a crater can be found for Edgar Allan Poe and another for Herbert George Wells.

What's more, I think a crater should be named for the late, great science writer, Willy Ley, who, more than anyone else in the world, managed to make ordinary mankind rocket conscious. He died three weeks before the lunar landing he had waited all his life to see, and surely there is a crater somewhere that belongs to him. (Such a crater has now been assigned.)

But now, with the indulgence of the Gentle Readers, I want to return to the subject of the opening of the essay.

On August 4, 1969, two weeks after the lunar landing and four days ago, as I write, my father died without undue or prolonged suffering, and after having been quite active, both physically and mentally, to the very end. I would like, in his honor, to tell another story about him.

My father always took the attitude that whatever worldly success and acclaim I achieved was simply what he expected of me and no more, and so he maintained a constant stern air of calm acceptance. To act visibly pleased would in his eyes (I imagine) have merely served to spoil me. (Behind my back, however, he was constantly praising me to all who would listen, and my mother, just as constantly, reported it all to me.)

Only once did this attitude falter. He had taken it for granted that I knew about science, but when I started publishing books on ancient history, he pulled me to one side and (looking furtively about as though he didn't want to be caught in this

* Yes I do, a little bit, and you know why.

act of weakness) said, 'Tell me, Isaac, how do you come to know all these things?'

And I said, 'I learned them from you, Pappa.'

He thought I was joking and I had to explain my meaning to him, as I will now explain it to you.

My father came to this country, in adult life, with no formal education in the ordinary sense (though he was extremely learned in Talmudic lore). He could never as much as help me with my homework.

What he could do, however, and what he *did* do, was to instill in me (and in my brother) a love of learning and a delight in explaining so firmly and so deeply that there was never any danger of losing it – and everything else followed quite automatically, and without particular credit to myself.

This avidity for learning and explaining has, as it happens, bought me a measure of material success – but quite aside from that, it has brought me an enrichment of life in a hundred ways beyond anything which can be measured by money or by any other palpable standard.

Thank you, Pappa.

4

Worlds in Confusion

In my recent book on the Bible,* I naturally had occasion to refer to the plagues that visited Egypt in the time of Moses, as described in the Book of Exodus. In doing so I said:

'Although these plagues, if they had taken place as described in the Bible, must have loomed large in any contemporary records or in later histories, no reference to them is to be found in any source outside the Bible. In 1950, Immanuel Velikovsky, in his book *Worlds in Collision* attempted to account for the plagues (and for some other events described in the Bible) by supposing that the planet Venus had undergone a near collision with the earth. The book created a moderate sensation among the general public for a while, but the reaction of astronomers varied from amusement to anger, and the Velikovskian theory has never, for one moment, been taken seriously either by scientists or by Biblical scholars.'

That's all I said, and it seems to me that I spoke gently and without undue heat. Nevertheless, the vials of wrath were opened upon me and I received a number of letters from ardent Velikovskians denouncing my innocent statement with a great deal of emotional fervor.

Which just goes to bear out my feeling that there is no belief, however foolish, that will not gather its faithful adherents who will defend it to the death.

It isn't even very difficult to see why Velikovskianism should be attractive among certain groups.

Velikovsky uses his theories to try to show that certain of the miracle tales in the Bible (Old Testament only, by the way) are more or less true. To be sure, he removes those events from the miraculous by taking away the hand of God and substitut-

* *Asimov's Guide to the Bible; Volume One, The Old Testament* (Doubleday, 1968).

ing a set of weird natural phenomena instead, but that makes no difference.

Velikovsky's book made the headlines as the work of a 'scientist' (which Velikovsky is not). It was ballyhooed as demonstrating that 'science' was proving the Bible true – though the amount of real science in the book could be placed in the eye of a needle without making it any more difficult to thread it.

Still, to all those who were brought up with traditional beliefs concerning the Bible, it was a great relief that science (the great enemy) had finally 'proved' all those miracles, and the book became a best seller.

Secondly, Velikovsky's views tended to make orthodox astronomers look foolish. Imagine those stupid professors not seeing all those things that Velikovsky presented so plainly!

There is always something pleasant about seeing any portion of the Establishment come a cropper, and the scientific Establishment in particular. Scientists, these days, are so influential, so far out of the ordinary clay, so supreme in their self-confidence, and (to put it in a nub) so 'smarty-pants' that it is a particular pleasure to see them stub their toe and go flat on their faces.

Those who experience the pleasure most acutely (I suspect) are those scholars who are not, themselves, scientists.

After all there was a time, one short generation ago, when the humanists were the scholars par excellence and the scientists were grubby fellows who worked with their hands and weren't up on Dostoevski. Now it is the scientists who hold top rank as scholars, who are most influential, most listened to, most honored, and (heavens!) get showered with government funds.

Naturally, many humanists find themselves attracted to a thesis that makes fools of scientists. And when some astronomers petulantly overreacted to the attention Velikovsky was getting and wrong-headedly tried to suppress him, the scholars were jubilant. They knew just enough science to cast Velikovsky in the role of a suffering Galileo.

Ever since, *some* non-scientist scholars have been lionizing

Velikovsky. It's not surprising that the most vehement letter I received was from an English teacher.

I don't want to beat a dead horse, but some of my correspondents self-righteously demand that I *read* Velikovsky before I denounce him. The implication is that if I only read him the truth of what he has to say will be borne irresistibly down upon me.

But, as it happens, I *have* read him, and I remain untouched. In fact, I think that if anyone reads *Worlds in Collision* and thinks for one moment that there is something to it, he reveals himself to be a scientific illiterate.

This is not to say that some of Velikovsky's 'predictions' haven't proved to be so. He predicted Venus would be hotter than astronomers suspected in 1950 and he was right. However, any set of nonsense syllables placed in random order will make words now and then, and if anyone wants to take credit for Velikovsky's lucky hits, they had better try to explain the hundreds of places where he shows himself not only wrong but nonsensical.

Thus, at the very beginning of the book, Velikovsky describes various theories as to the origin of the solar system and the development of the Earth. He stresses the shortcomings and insufficiencies of these theories, naturally, for he plans to advance a far better one himself. He says, on page 11:

'According to all existing theories, the angular velocity of the revolution of a satellite must be slower than the velocity of rotation of its parent. But the inner satellite of Mars revolves more rapidly than Mars rotates.'*

That's a very pretty paragraph but it is quite wrong, and it shows that as an astronomer, Velikovsky may quite possibly be an excellent psychoanalyst.

There is absolutely nothing in any astronomic theory I have ever heard of that relates the angular velocity of a satellite to the period of rotation of the planet it circles. Nothing requires that a satellite revolve about its planet either faster or slower than the planet's period of rotation.

* If you want to check my quotations, I am using the first edition of *Worlds in Collision*, Macmillan, 1950.

The angular velocity of a satellite depends on two things and *only* two things: the masses of its primary and itself, and the distance between the two bodies. If the primary is much larger than the satellite (as is usually the case), the mass of the satellite can be ignored.

The closer a satellite is to its primary, the more rapidly it moves in its orbit. If it is close enough to its primary, it will revolve about that primary in exactly the same time that the primary rotates, and if it moves still closer, then it will revolve about the primary in less time than it takes the primary to rotate.

'The inner satellite of Mars revolves more rapidly than Mars rotates' (to quote Velikovsky again) only because it is close enough to Mars to do so. At the distance of that satellite to Mars, it *can't* revolve any more slowly if Newton's law of gravity is to be obeyed. Far from defying 'all existing theories' by revolving so quickly, the inner satellite would defy them if it did *not* revolve so quickly.

Nor is Mars's inner satellite unique. The inner portions of Saturn's rings consist of innumerable tiny satellites, all of which revolve about Saturn faster than Saturn rotates about its axis. What's more, almost every artificial satellite man has launched into space has circled Earth in less time than it takes Earth to rotate on its axis. Earth rotates in 24 hours and some satellites have made the trip in 1.5 hours.

On page 49, Velikovsky writes:

'Ipuwer, the Egyptian eyewitness of the catastrophe, wrote his lament on papyrus: "The river is blood," and this corresponds with the Book of Exodus (7:20): "All the waters that were in the river were turned to blood." The author of the papyrus also wrote: "Plague is throughout the land. Blood is everywhere," and this, too, corresponds with the Book of Exodus (7:21): "There was blood throughout all the land of Egypt." '

I quote this because in my own book on the Bible I said there was no reference to the plagues of Egypt (which is the 'catastrophe' to which Velikovsky here refers) in any source outside the Bible, and one nice lady wrote me a letter in which she was most wroth at my having made that statement. She quoted Ipuwer, too, as an example of an outside source.

Who was Ipuwer? He was the author of a papyrus which has been dated back to the time of the Sixth Dynasty, about 2200 B.C. It was a time when the 'Old Kingdom' (which had built the pyramids) was in decay, and when Egyptian society was breaking down into feudalism, confusion and misery. Ipuwer didn't like the situation and described it very much in the tones with which Tacitus described the decaying Roman society of his time and the New Left describes the decaying American society of our own time.

And what does this have to do with the plagues of Egypt? Assuming the plagues took place at all, they took place at the time of the Exodus; and assuming the Exodus took place at all, it took place in the reign following that of Rameses II – about 1200 B.C.

In other words, Ipuwer's description was written a thousand years before the events Velikovsky claims it was describing. However, Velikovsky adjusts the dates. He shoves the Exodus backward from 1200 B.C. to 1500 B.C. and Ipuwer he shoves forward from 2200 B.C. to 1500 B.C. In a greater miracle than any in Exodus, he thus brings them together.

Of course, it may be that the Egyptologists are as wrong in their chronology as the astronomers are in their celestial mechanics.

Yet even if we accept Velikovsky's arbitrary dates, what are we to make of Ipuwer's words, which go on and on in their wailing? Is it possible, is it just barely possible, that he was making use of metaphor? If I were to say that 'Society is going to the dogs' would Velikovsky be justified in supposing that I was speaking of a band of wild dogs who had entered my city and were devouring its inhabitants?

In fact, Velikovsky depends throughout his book on the denial of metaphor. He quotes heavily from myths and legends of all nations, taking every word literally, treating them as though they were architect's blueprints. To be specific, he makes frequent use of passages from *Legends of the Jews* by Louis Ginzberg. I happen to have read Ginzberg (I wonder how many Velikovskians have?) and it would take a man chemically free of any trace of humor to take those medieval rabbinical tales seriously.

Of course, Velikovsky must be very selective. The entire corpus of humanity's myth and legend yields sentences on every side of every question and sometimes one of them must be hammered a bit to make it fit. Velikovsky talks about Atlantis, for instance, on page 147, saying:

'Critias the younger remembered having been told that the catastrophe which befell Atlantis happened 9000 years before. There is one zero too many here.'

So he removes it. What's a zero? Velikovsky makes the Atlantis catastrophe nine hundred years before Critias and now it fits his own chronology.

Gentle Reader, place all the myths and legends of the human race at my disposal; give me leave to choose those which I want to use and allow me to make changes where necessary; and I will undertake to prove anything you wish proven.

What Velikovsky is trying to prove is that a great comet was spewed out by Jupiter; that it kept ping-ponging back and forth, sideswiping Earth every once in a while, bringing about all the plagues of Egypt, stopping the Earth in time for Joshua's battle (the one in which he ordered the Sun and Moon to stand still) and so on.

After the comet was all through playing games, it settled down to become the planet Venus.

The scenes of impossibility this presents are simply colossal in their grandeur. If Venus emerges from Jupiter and passes the near neighborhood of Earth, it must have been in an extremely elongated and comet-like orbit to begin with. (The design of the solar system and the position of the planets make this necessary.)

For it then to settle down to a nearly circular orbit far, far away from Jupiter – such as it possesses today – cannot possibly be explained by any of the laws of nature worked out by scientists.

Nor does Velikovsky explain it. He simply says it happened. And his evidence? Well, there is his carefully selected list of sentences from myths and legends. And there is also the tale of an analogous event in the heavens.

On page 78, he says:

'That a comet, encountering a planet, can become entangled and drawn away from its own path, forced into a new course, and finally liberated from the influence of the planet is proved by the case of Lexell's comet, which in 1767 was captured by Jupiter and its moons. Not until 1779 did it free itself from this entanglement.'

Notice Velikovsky's use of words: 'entangled,' 'forced,' 'free itself' and so on.

I could scarcely blame any innocent who comes upon this passage if he believes that what happened was that Lexell's comet was trapped by Jupiter and his moons; that the comet was bounced from one moon to the next; that it was dribbled off Jupiter; that for twelve years it vainly tried to escape from this 'entanglement.' Then, finally, it broke away. Apparently it lowered its head, charged those nasty moons and plunged through at last.

What *really* happened was that Lexell's comet passed through the Jovian system in May 1767. It passed *through* it and did not linger; it was not entangled. Its orbit was changed as a result of gravitational attraction strictly according to astronomical laws of celestial mechanics. In its new orbit it passed near Jupiter again in the summer of 1779 and again its orbit was changed.

This happens frequently to comets.

Well, then, could the same thing have happened to Earth and comet Venus? Not at all! The situation is different!

Jupiter is an immense world, 318 times as massive as Earth. Lexell's comet is only 1/5000 as massive as Earth, at most. The orbits of Jupiter and its moons were not affected at all by the influence of the tiny comet.

Comet Venus, on the other hand, is nearly as massive as Earth. If comet Venus passed so close to Earth that its orbit was changed to the present one possessed by Venus, then Earth's orbit would have been changed radically as well – and heaven knows what would have happened to the Moon.

In fact, if Earth's orbit had been such as to give it a climate capable of supporting life before the encounter, it would almost certainly have had a new orbit that would have made it a planet incapable of supporting life after the encounter.

But let's not think of such gigantic catastrophes. Let's not think of altered orbits, of oceans leaving their beds and slopping over the continents. Let's not think of the *great* results of Earth's suddenly stopping its rotation when Joshua commanded the Sun to stand still. (Not only would Joshua's soldiers all have fallen down and rolled for a thousand miles, but the energy of rotation would have been converted into heat and have melted the Earth's crust.)

Instead of all that, I'll just mention one little thing. There are many limestone caves in the world in which many stalactites and stalagmites have been slowly and precariously forming over a period of hundreds of thousands of years. They are quite brittle.

If the Earth had stopped its rotation at the time of the Exodus, or if it had even slightly changed its period of rotation, every one of those stalactites and stalagmites would have broken.

They did not! They are there! Intact and beautiful, as you will see for yourself if you visit any limestone cave. And those stalactites and stalagmites, standing there mutely, are stronger evidence against Velikovsky's theory than all Velikovsky's selected lines from myths and legends can possibly counter.

But let us move on. Velikovsky needs a rain of burning fire to explain certain Biblical allusions and he finds a great deal of talk about such combustive events in his myths.

You and I might suppose that the experience of a volcanic eruption is terrifying enough to account for such tales and can easily be magnified to a whole sky on fire, given the inevitability of poetic license. Velikovsky, however, does not believe in either poetry or metaphor. He wants a literal rain of fire and he uses comet Venus to explain it.

On page 53, he says:

'The tails of comets are composed mainly of carbon and hydrogen gases. Lacking oxygen, they do not burn in flight, but the inflammable gases, passing through an atmosphere containing oxygen, will be set on fire.'

These are impressive sentences. The very phrase 'carbon and hydrogen gases' takes my breath away. Hydrogen is, indeed, a gas at ordinary cometary temperatures, but carbon is *not*. It is, in fact, among the least gaseous substances known and it takes a temperature of 4200° C. (7500° F.) to make it gaseous.

Now I am a chemist. If Velikovsky wants to say that Laplace's analysis of celestial mechanics is all wrong and that Venus can emerge from Jupiter and settle down in its present orbit, I will smile. If he wants to say that Egyptologists don't know the difference between 1200 B.C. and 2200 B.C., I will grin.

But if he says carbon is a gas, *that's going too far*.

Let's not be too hard, though. As a matter of fact, the tails of comets seem to be made up, at least in part, of molecular fragments, some of which contain both carbon and hydrogen and are therefore 'hydrocarbon' in nature. It may be that this is what Velikovsky had in mind when he spoke of 'carbon and hydrogen gases.'

To be sure, this chemical analysis of comets' tails is the result of some very esoteric and sophisticated astronomical theories, and you might wonder how Velikovsky can come to accept them. After all, if astronomers are so far wrong on the simplest tenets of celestial mechanics, can they be trusted in the delicate nuances of spectroscopy? But then, the astronomical decision with regard to the chemical structure of comets' tails suits Velikovsky's theory, so he accepts it.

(Gentle Reader, give me the chance to pick and choose among the findings of science, accepting this and rejecting that according to my lordly whim, and I will undertake to prove anything you wish proven.)

But granted the hydrocarbons, can cometary tails really blaze up if they pass through Earth's atmosphere? Can they really cause rains of fire? No, sir, not a chance.

Those comet tails are just about the thinnest gas you can imagine. Some tails have extended outward through space for a hundred million miles, but if all that glowing next-to-nothingness were condensed to the thickness of ordinary gases in our own atmosphere, they would perhaps fill a living room or two.

You know what happens when the Earth passes through a comet's tail? Nothing!

How do I know? Because the Earth has passed through one on a number of occasions. It passed through the tail of Halley's Comet in 1910. Many people refused to believe the scientific Establishment who said nothing would happen. They thought the end of the world would come. Or they thought the poisons in the comet's tail (they believed the spectroscopic analysis) would kill all life on Earth.

And what happened? As I told you, nothing!

Of course, comet Venus was much huger than an ordinary comet. The ordinary comet has the mass of a small asteroid. Comet Venus had four fifths the mass of the Earth itself. Its 'tail' must have been much more voluminous than that of an ordinary comet. Could it be that comet Venus' atmosphere did ignite on passing through the Earth and did set up rains of fire?

And can Venus' atmosphere still be hydrocarbon in nature today? Velikovskians think it is. After Mariner II had made a close approach to Venus in 1962, someone at a news conference made an unfortunate remark that could be interpreted as indicating that Mariner II had indicated the atmosphere of Venus was indeed hydrocarbon.

The misinterpretation was corrected at once and has been corrected with the periodicity of a tolling bell ever since, but it doesn't help. The Velikovskians insist that science has now determined that Venus' atmosphere is hydrocarbon in nature and won't budge from that.

Actually, all the data that have been collected on Venus since Velikovsky's book was published make it seem more and more definite that its atmosphere is about 95 per cent carbon dioxide and perhaps 5 per cent nitrogen – as non-inflammable a gas mixture as you can imagine.

There are ice crystals in the upper atmosphere making up its clouds and a very recent suggestion is that there are small quantities of various mercury compounds vaporized by Venus' great heat and floating about in the lower atmosphere. Even so, the atmosphere would remain non-inflammable.

But suppose comet Venus' atmosphere *was* hydrocarbon. And suppose it did ignite spontaneously in the upper atmosphere and send down rains of fire. And suppose more of the

57

hydrocarbon remained aloft in dank, oily mists, darkening the surface of the Earth.

It is hard to see how such a supersmoggy situation could fail to kill everything on Earth, but never mind. Velikovsky needs it to explain the ninth plague of Egypt (darkness) and bears it out with the usual mishmash of statements drawn out of context from legends, myth and poetry.

But then what happened to those hydrocarbon clouds? Velikovsky explains that on page 134:

'When the air is overcharged with vapor, dew, rain, hail or snow falls. Most probably the atmosphere discharged its compounds, presumably of carbon and hydrogen, in a similar way.'

That sounds as though the hydrocarbon clouds precipitated in a rain of gasoline, kerosene and asphalt. Worse and worse!

But then Velikovsky says, 'Has any testimony been preserved that during the many years of gloom carbohydrates precipitated?'

Where has the word 'carbohydrates' come from? Can it be that Velikovsky doesn't know the difference between 'hydrocarbon' and 'carbohydrate'? Gasoline is an example of a hydrocarbon material; sugar is an example of a carbohydrate.

Can a cloud of gasoline vapor precipitate as a sugar-like compound? This, I'm afraid, is chemically impossible.

Can there have been sugar vapor in the clouds in the first place? This, too, I'm afraid, is chemically impossible.

But then why does Velikovsky seem to think there will be a precipitation of carbohydrate?

Ah, he has to explain the fall of manna, you see, the miraculous food on which the Israelites lived for forty years in the desert. To achieve the manna, all Velikovsky had to do was to talk of carbon and hydrogen and, at a key point, quietly slip in that word 'carbohydrate.' Presto, change, alakazam, and there you are.

All you need is an abysmal ignorance of chemistry and you're set.

But that's enough. My space is used up and why go further? At least fifty more passages can be chosen from the book but

I've proved my point, I think, that Velikovsky's theories are simply silly.

In fact, sooner than have the miracles of the Bible explained by such a farrago of broken astronomy, half physics and semi-chemistry, I would accept them exactly as given in the Bible. If I must choose between Immanuel Velikovsky and Cecil B. De Mille, give me De Mille, and quickly.

B
Physics

5

Two at a Time

Several days ago the telephone rang and a young male voice, having ascertained that I was indeed I, said, 'Pardon me, sir, how do you determine the center of gravity of the Earth-system?'

The speaker was most courteous and I recognize the fact that I have certain duties. If I'm going to act publicly as though I know everything and if I'm going to proceed to make my living out of that pretense, the least I can do is answer simple questions when those are put to me politely.

I said, 'The Earth is eighty-one times as massive as the Moon. That means if you draw a line between the center of the Earth and the center of the Moon, the center of gravity is on that line at a point eighty-one times as far from the Moon's center as from the Earth's center.'

'Oh,' he said, 'but how far above the Earth's surface would that be?'

'It wouldn't be,' I said. 'It's roughly a thousand miles *beneath* the Earth's surface.'

'Aha,' said my young friend, 'I knew he was trying to catch us.'

A pang of dismay clutched my heart. '*Who* was trying to catch you?'

'My teacher,' he said cheerfully. 'This is my homework.'

And he hung up.

So I give fair warning. No more question-answering by phone. I'm not going to be made an innocent party to cheating.

But all is not lost. It set me to thinking—

The fact that the center of gravity of the Earth–Moon system is located a thousand miles under the Earth's surface must not obscure the fact that it is located 2900 miles from the Earth's center in whatever direction the Moon happens to be at the moment. The Moon revolves about that center (which is at one of the foci of the Moon's orbital ellipse) and *so does the Earth!*

The center of the Earth makes a small monthly ellipse about the center of gravity of the Earth–Moon system, tracing out a curve precisely similar to the lunar ellipse, but only 1/81 its size.

The fact that the Earthly ellipse is so small does not matter. What is crucial is that the Earth moves in response to the Moon's gravitational pull just as freely as the Moon moves in response to the Earth's.

Indeed, every mass-possessing particle (I will leave out that adjective henceforth and assume it to be understood) in the Universe is the center of a gravitational field and every particle in the Universe moves freely in response to the gravitational field of every other particle (unless constrained by some other type of field).

Let's consider some very simple universes. A Universe with no particles can be dismissed as trivial. So can a Universe with one particle, for though that particle may originate a gravitational field, the field cannot be detected unless another particle is inserted into the Universe. What we detect is never the gravitational field in itself, but always a gravitational interaction.

So a Universe with two particles is the simplest we can consider.

If the two particles are at rest with respect to each other, they will interact, gravitationally, in such a way as to approach each other at an accelerating velocity until they collide. If they are already moving toward each other along the line connecting them, the same thing will happen.

If they are moving away from each other along the line connecting them, then they will move away from each other at a decelerating velocity. If the initial motion is less than escape velocity, the deceleration will eventually bring them to rest, after which they will begin to approach and finally collide. If the initial motion is more than escape velocity, they will decelerate at a slowly decreasing rate that will never bring them to mutual rest. They will move apart forever.

If they are moving, to begin with, in some direction that is *not* along the line connecting the two at a velocity less than escape, they will trace out a pair of interlocking ellipses (as Earth and Moon do). The two ellipses will be similar, with sizes that are inversely proportional to the masses of the particles. Depending on the velocity and masses the ellipses can be of any eccentricity from 0 (for a circle) to 1 (for a parabola).

If the two particles are moving relative to each other at a velocity greater than escape, they will each trace out a hyperbolic path and separate forever.

All these possibilities can be deduced with precision by means of a relatively simple equation first worked out by Isaac Newton nearly three centuries ago, and since modified, to take care of certain additional refinements, by Albert Einstein.

But suppose there are more than two particles in the Universe. In that case, each particle would move in response to the algebraic sum of all the other gravitational fields and this would constantly shift as every other particle would likewise move in response to the algebraic sum of the other gravitational fields.

For more than two particles there is no general equation that will exactly describe the motions of each – or at least none has been worked out. There is no general equation that will cover even the simple case of a Universe containing three particles. Three centuries after Newton, the so-called 'three-body problem' has not been solved.

In fact, it's even worse than this, for according to the strictest interpretation of Newton's law, a 'particle' is something that has mass, but zero volume, and nothing of the sort exists in the real world. Consequently, even the 'two-body problem,' which *is* solved, does not truly apply to the real world.

This sounds as though Newton's theory of gravitation is the purest fantasy. After all, if it works exactly only for two non-existent particles and for nothing beyond that, we might as well go back to that old bromide of medieval scholasticism and count the number of angels who can dance on the head of a pin. Or might we?

There is a difference, you see. Even if two clerical scholars agreed on the exact number of angels who could dance on the head of a pin, to what use could they put the result? Newton's theory, on the other hand, divorced from reality though it seems, can be brought down to Earth and put to work.

When we say the Three-Body Problem is not solved, we mean in terms of pure mathematics. No definitive and totally precise answer can be given in terms of a finite and universally applicable equation.

The astronomer, however, working on celestial mechanics itself and not on the pure mathematics analogous to it, asks not for a system describing the motion of various bodies with ultimate exactness (though he would have no objection to one), but will be satisfied with a system that describes positions and motions within the limits of error of observation for some reasonable period of time. In other words, he will be satisfied to work with useful approximations.

Let us begin, then, with a subatomic particle. It has mass and yet it has a volume which, while not actually zero, is *nearly* zero. It is *almost* a point source of a gravitational field that will serve the Newtonian ideal. The only trouble with these subatomic particles is that their masses are so small that their gravitational interactions are just about undetectable, particularly since such particles are likely to undergo other interactions, involving fields which are much, *much* stronger than the gravitational.

Gravitation, you see, is one of four types of interaction known in the Universe. Two of these, however, are confined to atomic nuclei and can be ignored if we are dealing with particles as large as an atom.

The remaining interaction is electromagnetic. It is the dominant interaction in all objects ranging from the size of an atom to that of a small asteroid. The forces holding solids in one piece are electromagnetic in nature.

The electromagnetic interaction is enormously more intense than the gravitational. A small asteroid with an irregular shape is easily held in that shape by electromagnetic interactions, even while the gravitational interactions of the particles making it up are trying to force it into a spherical shape. The small asteroid has no trouble maintaining its electromagnetic-supported irregularity against gravitational demands.

The electromagnetic interaction, however, has both attractions and repulsions, and in sizable chunks of matter these usually balance in such a way that the excess, one way or the other, is quite small. The gravitational interaction, however, involves attractions only (as far as we know) and as an object increases in mass, in density or in both, the total intensity of its gravitational field increases without theoretical limit. For large asteroids, and certainly for bodies the size of the Moon and the Earth, a spherical shape is a necessity. The gravitational field insists upon it and at those masses it cannot be denied.

But how can we be sure of that? In bodies the size of the Moon or the Earth there are trillions of trillions of trillions of particles in close contact. Each particle in the Earth must interact gravitationally with every other particle. What's more, each particle in the Earth must interact gravitationally with each particle in the Moon.

Can a theory which can't deal with total effectiveness with even three particles be of any use whatever when trillions of trillions of trillions are involved?

When Newton was working out his theory of gravitation, this stopped him. He thought he knew the answer, but he couldn't prove it. In order to prove his intuitively expected answer to be true, he needed powerful mathematical tools that did not yet exist. Fortunately, he was Newton. He invented the necessary mathematical tool himself – calculus.

Using calculus, Newton was able to show that *if* a real astronomical body were (1) spherical and (2) either of uniform density throughout or of density that changed equivalently in all directions from the center, *then* the body produced a gravitational field exactly like that which would be produced by a body of similar mass but no volume (a point source) located at the center of the body.

The Earth, for instance, is roughly spherical, and we have every reason to suppose that its density varies (when passing from center to surface) in precisely the same way no matter which direction we choose. Therefore, the gravitational field seems to originate at the Earth's center and no matter where we are located on Earth's surface, we are pulled toward that center.

The Moon also appears reasonably spherical, as does the Sun and those planets whose discs we can make out in a telescope. It is fair to suppose that, with occasional special exceptions, all astronomical bodies fulfill the criteria of shape and density distribution, and all can be treated as though they possessed point sources of gravitational fields at their center.

Of course, if you're going to pretend an astronomical body is a point source, you must treat it as one. You can't get inside a point source since it has no volume to get inside. Therefore, if you tunnel into the body of the Earth, the Newtonian equation in its simplest form breaks down.

If there were really a point source of gravitation at the center of the Earth, then the closer and closer you dig your way through the Earth toward that center (or find your way there through long caves, à la Jules Verne) the more and more intense would be the gravitational interaction to which you are subjected. Finally, when you reach the very center, the intensity of interaction (provided you are a point source yourself) becomes infinite.

In actual fact, the closer you dig or probe your way toward the center of the Earth, the weaker the gravitational interaction. And if there were a hollow at the very center of the Earth (à la Edgar Rice Burroughs), the gravitational interaction within that hollow would be zero everywhere no matter how small or large it might be (something ERB didn't know).

This is by no means a 'paradox' which somehow goes against the theory of gravitation. It merely falls outside the original simplifying assumption of point sources. If you divide the Earth into two portions, the part closer to the center than yourself and the part farther from the center than yourself, or if you divide it into the hollow and the part outside the hollow, you can show that the Newtonian equation will explain the situation.

There are other difficulties, too. The Earth is *nearly* spherical, but it is not *exactly* so. It is actually an oblate spheroid, meaning that the distance from the center to the surface varies in this fashion: it is at a minimum to the North (or South

Pole) and increases as one approaches the equator (in any direction) reaching a maximum at the equator. The equatorial radius is 13 miles greater than the polar radius; not much in a total radius of some 3950 miles, but enough.

This bulge produces a tiny gravitational interaction of its own with the Moon, and it is this which causes the ends of the Earth's axis to mark out a slow circle that takes 25,780 years to complete. This 'precession of the equinoxes' would *not* take place if the Earth were a perfect sphere.

These departures from maximum simplicity may complicate the handling of Newton's law of universal gravitation but they also strengthen it. The fact of the precession of the equinoxes was observed two thousand years or more before Newton, but the rational explanation of that fact had to wait for Newton's equation, properly applied.

In fact, the Earth is not even a perfect oblate spheroid but is rather lumpy and uneven everywhere. Its land surface is jagged with mountains and depressions and the density of the outer crust shows irregular patches of high and low values. As methods for detecting the intensity of the gravitational field with greater precision are developed, tiny differences are detected from spot to spot. These reflect the departures of the Earth from the Newtonian ideal.

And yet these departures are indeed tiny departures. It is not necessary for us to begin with the Earth in its *exact* shape and with its *exact* density distribution and try to work out the *exact* nature of its gravitational field. If we tried to do this, the problem would be far too complicated and we would surely fail.

Instead, it is only rational to begin with an idealization, a simplification, even though we know that to be 'wrong.' We begin with that as a first approximation, then correct for major discrepancies, then for minor discrepancies, then for very minor discrepancies and so on. Little by little we approach a (possibly unattainable) real 'truth' and in the process develop a precision as tight as necessary for our purposes.

Undoubtedly, bodies other than the Earth possess these same imperfections; these same minor departures from perfect sphericity and perfect density symmetry.

Consider the Moon. The Moon rotates very slowly compared to Earth and it is therefore a much closer approximation of a sphere. There is no equatorial bulge to speak of.

Yet if the Moon were entirely perfect in its Newtonian simplicity, an object in orbit about it would be expected to move in a certain fashion that could be calculated out to a considerable number of decimal places. This proved not to be so.

Lunar orbital satellites circling the Moon moved a little too quickly in certain spots in its orbit. All the known factors were included but there was still something left over. In order to explain the discrepancy, it was necessary to postulate that the Moon's gravitational field was a trifle more intense over certain sections of its surface than over others.

It was more intense, apparently, over the comparatively flat and unruffled 'seas' than over the cratered and mountainous areas. It seemed that under the seas were 'mass concentrations,' regions of higher-than-normal density. The term mass concentrations was quickly abbreviated to 'mascons' and this is the new magic word in selenography.

What are the mascons? One possibility seems particularly attractive. Suppose the seas are the sites where great meteors struck the Moon at some late stage of its development. If that be so, might there not be large meteoric lumps underlying the surface of the seas. If the lumps are largely iron, they would be twice as dense as the ordinary crust of the Moon. That would account for the tiny gravitational anomaly.

Astronomical objects which are so small that the gravitational field has not become the overriding factor may maintain gross irregularities of shape and be nothing even approaching the spherical. The asteroid Eros is a notorious example, for it is, apparently, brick-shaped, with its longest axis about fifteen miles in length.

This means that its gravitational field in its own near vicinity varies in complicated fashion from place to place. The intensity of the gravitational field of such a body is very weak, however, and if you were standing on Eros' surface you would be subjected to a gravitational interaction only about one thousandth that of the Earth.

It is that high only because on the surface you would be standing but a few miles from its center. If you were 1000 miles from Eros' center (as you are 4000 miles from Earth's center when you stand on Earth's surface), Eros' gravitational interaction with you would be more like a billionth of what you are used to on Earth.

This is and must be true of any astronomical body small enough to be able to deviate widely from the spherical. The intensity of its gravitational interactions are tiny and play little, if any, role in astronomical calculations. At a distance, besides, any variations in so tiny a gravitational field are even less important than the field itself and Eros and other objects of that nature can be regarded as point sources anyway, provided we are not immediately on, or very near, its surface.

Even if we pretend that all astronomical bodies behave as point sources, there still remains the question of the Three-Body Problem. How can we predict the movements of the Moon, for instance, in a Universe made up of innumerable objects all producing gravitational fields, even if each object, including the Moon, is considered a point source?

Fortunately, the distribution of bodies in the Universe is such that there is always good reason to consider them two at a time, at least to begin with. When a third body is present it is always so small it may be ignored, or so distant that the first two bodies can be considered as a single point source. Either way, we are left with a Two-Body Problem.

Suppose we consider the Moon and the Earth. These two bodies are separated by a distance of 237,000 miles (on the average) and there is no other sizable body closer than a hundred times that distance. As a first approximation, then, we can pretend that the Moon and the Earth are alone in the Universe and treat them as though they represented a Two-Body Problem.

When this is done, it follows that the Moon and Earth travel in a pair of interlocking ellipses about the center of gravity of the system. The Earth's ellipse is so small that it can be disregarded, at least in lay discussions, and we can say, 'The Moon revolves about the Earth,' without being corrected, even by astronomers.

It is from the relative sizes of these ellipses that one can deduce that the Earth has eighty-one times the mass of the Moon.

The Earth–Moon system is 93,000,000 miles from the Sun. There are other bodies closer (Mercury, Venus and Mars, when they are on the Earthward sides of their own orbits). The Sun, however, is over 150,000 times as massive as all the inner planets put together, so that the Earth–Moon system (taken as a point source at the center of gravity) and the Sun can be treated once again as a Two-Body Problem.

When that is done, it turns out that the center of mass of the Earth–Moon system moves around the Sun in an elliptical orbit (not very far removed from a circular one) in $365\frac{1}{4}$ days. To be very precise, the Earth–Moon system moves about the center of gravity of the Earth–Moon–Sun system, which is about three hundred miles from the center of the Sun. The center of the Sun makes a tiny little ellipse about that center every year – or would, if Earth, Moon and Sun were the only bodies in the solar system.

Both Earth and Moon revolve about their own center of gravity twelve times and a little over each time the system revolves about the Sun. This means that the orbit traced out by the Moon's center marks out twelve very shallow waves (and the beginning of a thirteenth) as it moves around the Sun. The Earth's center marks out a similar set of waves but considerably shallower.

By comparing the effect on the Earth of the Sun at its known distance and the Moon at its known distance, it is possible to determine that the Sun is 27,000,000 times as massive as the Moon and therefore 330,000 times as massive as the Earth.

Of course, the Moon responds also to the gravitational fields of Earth's equatorial bulge, as well as those of Venus, Mercury, Mars, Jupiter and so on. The intensity of these various gravitational interactions vary constantly as Moon, Earth, Venus and all the rest move in their respective orbits at speeds which are not entirely constant.

All these other gravitational interactions introduce only minor corrections ('perturbations') in the Moon's orbit, and its general shape as worked out by two-body considerations is not seriously altered. Nevertheless, astronomical precision requires that these other interactions be taken into account. I am told that the equation representing the motion of the Moon with all the known perturbations taken into account would fill a large volume and would still be only an approximation, though a very close one. Newton was reported to have said that working out an equation for the motion of the Moon was the only problem that ever made his head ache.

What about other planet–satellite systems? Jupiter has twelve known satellites of which four are roughly the size of our Moon. Jupiter itself is so much more massive than all its satellites put together, however, that each satellite can be studied in connection with Jupiter alone as a two-body problem.

If we know the distance of a particular satellite from Jupiter and the time of its revolution about that planet, we can compare that time of revolution with the time it would take to revolve about Earth at the same distance from ourselves. The satellites whip about Jupiter much more quickly than they would about the Earth and from that, making use of Newton's equation, we can calculate the intensity of Jupiter's gravitational field relative to Earth's and, therefore, its mass as well. It turns out Jupiter has a mass 318 times that of the Earth.

Similar calculations can be made easily for any planet with a satellite whose distance from the planet and period of revolution can be determined.

But how about the masses of the satellites themselves? The mass of the Moon was easily determined from its effect on the motion of the Earth. Alas, this is an unusual case. The mass of the Moon is so large a fraction of the mass of the Earth that the Moon makes the Earth wobble noticeably. This is not so for any other satellite in the solar system. One and all they are so small in proportion to their planets that their effect on that planet's motion is unnoticeable.

The mass of a particular satellite of Jupiter must be deduced from the perturbations it produces on the orbits of the other satellites and such calculations are not very precise.

Similar imprecision attends the masses of planets without satellites. Until recently, the mass of Venus had to be calculated from its perturbations on the Earth–Moon system, and the best one could do was to say that the mass of Venus was 0.8 that of the Earth. Now, however, we have sent Venus probes skittering closely past the planet, and from the effect of Venus on those probes, it has been calculated that Venus has a mass 0.81485 times that of Earth.

As you see, though, all mass determinations based on orbital movements are relative. They all boil down to a certain multiple of the Earth's mass, for instance.

To make all those relative determinations absolute, the absolute mass of one astronomical body, at least, must be determined, and that, indeed, was done.

The mass so determined was that of the Earth itself; the time was 1798, the place was England, the person was Henry Cavendish, the discussion thereof – well, please be patient.

6

On Throwing a Ball

Several years ago, I delivered a talk at my old alma mater, Columbia University, which went very well indeed, I am glad to say, and after it was over, some of the students in the audience came up to give me a very special present.

It was a sweat shirt. On the front of it was a picture of Isaac Newton with his name in bold letters underneath. On the back, moreover, was the eloquent legend: $f = ma$.

You can well imagine that I was delighted to receive it and that I wear it at every appropriate opportunity.

To be sure, I don't have quite as many opportunities to do so as a teen-ager would. At my advanced age (somewhat over thirty, as I've said before), my social engagements tend to be of a type at which colorful sweat shirts are frowned upon.

Yet I manage. Every once in a while I wear it, and when I do, the rest of the party suffers agonies, for I attract attention. I'm not aware of this myself, you understand, for as a result of a long lifetime spent in strenuous and nearly exclusive preoccupation with what goes on inside my skull, I have learned to be oblivious to the outside world. A small matter like being followed by puzzled, whispering teen-agers leaves me untouched.*

Mostly, what seems to attract their attention is not the picture of Isaac Newton (who, to them, is clearly a rock-and-roll sensation since he has long hair) but the mystic legend on the back. I imagine they try to work out its meanings and

---

* Actually, this is by no means the worst sweat shirt I own. Half a year ago, one of the many beautiful girls at Doubleday & Company presented me with a sweat shirt on which was written in bold, white block letters: ISAAC ASIMOV IS A GENIUS. I am ashamed to admit that I don't quite have the guts to wear it in public. I have always thought that my immodesty was limitless, but apparently it isn't.

speculate on its possible obscene significance (the times being what they are).

So why not explain it here?

Let's begin by throwing a ball. The ball is motionless when you start moving your arm, but by the time it leaves your hand it is traveling at a respectable speed. In the time that you were engaged in the act of throwing it, it was gaining speed from zero at the start to whatever amount it had when it sailed out of your clutching fingers. Such a gain of speed is called an 'acceleration.'

(It is better, actually, to speak of 'velocity' instead of speed. Velocity is a combination of speed and direction; if you speak of 'constant velocity' you mean motion at a constant speed *and* in an unchanging direction. Any change in velocity, whether involving a change in speed, a change in direction or both, is an acceleration.)

But in accelerating the ball, we have had to make an effort. The ball will not accelerate without one. It will not suddenly, all by itself, stop being motionless and begin to go faster and faster. We have to *throw* the darn thing to make it do so. What's more, the effort we make has to be applied to the ball. We can make all the throwing motions we want, but if the ball is lying ten feet away in the grass while we are doing so, nothing will happen to the ball. What's more, when we do throw the ball, that ball accelerates in the direction we are throwing.

Making an effort and applying that effort directly to an object is to exert a 'force' on that object, so what we are saying is: An object can be made to accelerate if, and only if, a force is exerted upon it, and the acceleration so produced is in the direction of the force.

A statement like that is sometimes called a 'law of nature' but that always strikes me as too portentous a title. It is simply a generalization. It is the common experience of mankind that acceleration and forces go together.

But back to the ball! If we throw harder, the ball is made to go faster at the time it leaves our hand. The change in velocity while we were throwing is greater. In short, the greater the force, the greater the acceleration. Again, this is the common experience of mankind.

In fact, when physicists began to measure force and acceleration with precision, they found that if exactly twice the force was applied to a particular body, then exactly twice the acceleration was achieved*; if exactly n times the force, then exactly n times the acceleration.

A short way of saying this is: Acceleration is directly proportional to force.

An even shorter way is to make use of mathematical symbols. Let acceleration be represented by a and force by (what else?) f. To represent direct proportionality, we make use of a wiggly mark, $\sim$. So we write:

$$a \sim f \qquad \text{(Equation 1)}$$

Let's pass on. What if we try throwing different objects? Suppose we throw a tennis ball with a certain amount of effort, then (to the best of our judgment) use just the same effort to throw, in succession, a baseball, a softball, and a shot (one of those metal spheres that shot-putters love to heave).

You know that using the same force you will not be able to make the baseball go as fast as the tennis ball. The softball will go slower still and the shot will hardly move.

It is the common experience of mankind, then, that a given force will accelerate a heavy object *less* than it will a light object. In fact, if you make measurements you will find that if x is twice as heavy as y, then a given force will accelerate x just half as much as y; if x is three times as heavy as y, x will be accelerated just a third as much as y and so on.

You might argue this point at once, maintaining that if this were so then a feather, much lighter than a baseball, ought to be accelerated correspondingly more, so that with equal effort we could make a feather move much more rapidly than a baseball. And we know that's not so. Nothing we do will make a feather move quickly.

But then our throwing arm is not the only force upon the

---

* Well, not *quite* exactly. Einstein's relativity introduces a correction that is vanishingly small under ordinary circumstances, but this essay is devoted to the Newtonian approximation and I am ignoring, for now, the better Einsteinian approximation.

feather. Air resistance sets up a force in the direction opposite to the one we are exerting with our arm. For reasons we need not go into, this counterforce is much more effective on a light object such as a feather than on a relatively heavy one such as a baseball. It's the *net* force that counts; the force that is left over after all forces are taken into account that controls the acceleration.

Again, if you try to push a very heavy object across a floor you may think a small acceleration ought to be produced; yet instead, there *is none at all*. The object won't budge no matter how long and steadily you push. This time there is a frictional force countering the one you exert and this one is more effective where heavy objects are concerned than light ones (all other things being equal).

In short, real life is rather complicated and that is why it took a couple of millennia of hard thought before some apparently simple generalizations of motion were worked out. It took transcendent genius to cut away the extraneous complications.

If we now ignore those extraneous complications and leave them out of account, we can say that the heavier an object is, the less acceleration a given force will induce in it.

But let us not say 'heavier' because this will raise complications. Let us instead invent a word 'mass,' which we can define as follows: 'Mass is a property of a body which affects the acceleration induced on that body by a given force in such a way that the greater the mass, the correspondingly smaller the acceleration.' (It turns out that under ordinary circumstances, the phrase 'more massive' is roughly equivalent to what we mean when we say 'heavier'; 'less massive' to 'lighter.')

The greater the mass of a body, the more difficult it is to accelerate it; that is, to alter its velocity. The resistance to change in velocity is spoken of as 'inertia.' This means, then, that the greater the mass of a body the greater its inertia. In fact, by definition, mass and inertia are different names for exactly the same property.*

* The late E. E. Smith, in his classic 'Lensman' tales postulated an inertia-free drive to get around the speed-of-light limit for space travel. Ordinary mass with inertia cannot go faster than light, he

If, then, for a given force, acceleration gets smaller as mass gets larger, we can say that acceleration is inversely proportional to mass.

To see how that might be presented in mathematical shorthand, let's represent mass as m and consider the quantity $1/m$. As m gets larger: 2, 3, 4, 5 and so on, $1/m$ gets smaller: $\frac{1}{2}, \frac{1}{3}, \frac{1}{4}, \frac{1}{5}$ and so on. In fact, $1/m$ gets smaller as m gets larger in exactly the same way that acceleration gets smaller as m gets larger.

If two variables are each inversely proportional to a third variable, then the two variables are directly proportional to each other. By which I mean that if acceleration and $1/m$ are both inversely proportional to m, then acceleration is directly proportional to $1/m$. We can say, then, that:

$$a \sim 1/m \qquad \text{(Equation 2)}$$

If acceleration is directly proportional to each of two different quantities, then it is directly proportional to the product of those quantities. In other words if a is directly proportional to f and to $1/m$ (see Equations 1 and 2) then it is directly proportional to $f \times 1/m$. We can say, then, that:

$$a \sim f/m \qquad \text{(Equation 3)}$$

When two properties are related by a direct proportionality, it means that as one grows larger (or smaller) the other grows correspondingly larger (or smaller). Yet one might be consistently twice as large as the other, or five times as large or 1.752 times as large. However much both grow larger or smaller in perfect step, the ratio of size remains the same. It stays 2 or 5 or 1.752 or whatever it is.

In order to change a direct proportionality to an equality, therefore, one might discover what that constant ratio is and multiply the appropriate side of the equation by that ratio.

If we don't happen to know just what the ratio is, in a particular case, we can just give it the general name of 'proportionality

---

suggested, but mass without inertia can go at any velocity, however great. It was a fascinating suggestion and I loved it, but if we look at it in the hard light of reality, we must admit that mass without inertia is equivalent to mass without mass – a contradiction in terms. (At least, so it seems.)

constant' and symbolize it, usually but not always, as k. (Why k? That was adopted from the Germans, who spell constant 'Konstant.')

If, then, we multiply the right-hand side of Equation 3 by such a proportionality constant, we establish an equality and we can write:

$$a = kf/m \qquad \text{(Equation 4)}$$

The presence of the proportionality constant is a pain in the neck, and physicists do their best to get rid of it by some legitimate means. In this case, we can choose the units of acceleration, force and mass in such an interconnected way that k will work out to be unity. And, of course, when a number of terms are multiplied together and one of them has a value of 1, it doesn't affect the product and can be omitted. *Provided we keep the units properly interconnected*, we can write Equation 4 as:

$$a = f/m \qquad \text{(Equation 5)}$$

By simple algebraic manipulation, Equation 5 can be converted into:

$$f = ma \qquad \text{(Equation 6)}$$

and that (*aha-a-a-a-a-a!!*) is what is on the back of my sweat shirt.

The connection with Newton is not hard to explain. All this stuff that seems so simple *now* is only simple because Isaac Newton explained it first in his book *Principia Mathematica*, published in 1687. What I present you with in Equation 6 (and on the back of my sweat shirt) is the simplest expression of Newton's Second Law of Motion.

Why Second Law? Because there is a First Law.

One way of stating Newton's First Law of Motion is this: 'If a body is not acted on by a force, it will remain at rest, or if it is already in motion, it will maintain a constant velocity, changing neither its speed nor its direction of travel.'

A constant velocity implies zero acceleration. Newton's First Law of Motion therefore reads, using mathematical shorthand, 'If $f = 0$, then $a = 0$.'

But if we look at Equation 6, we can see (assuming a body has *some mass*) that if $f = 0$, then a *must* equal 0:

$$0 = m \times 0 \qquad \text{(Equation 7)}$$

It turns out, then, that Newton's First Law of Motion is merely a special case of Newton's Second Law, and Equation 6 is an adequate expression of both the First *and* Second Law of Motion. (Why bother with a First Law then? Couldn't Newton see it was mathematically unnecessary? Sure, he could. The thing was, though, that he was busy establishing a new world picture and he had to knock down the old world picture first. The First Law destroyed the keystone of the old system and so it was psychologically necessary to present it first.)

There is also a Third Law of Motion, also advanced by Newton, which states that if Body A exerts a certain force on Body B, then Body B exerts an equal force (but in the opposite direction) on Body A.

This is usually called the 'Law of Action and Reaction' and if ever there was an unfortunate name, that's it. It gives an utterly false impression that has confused innumerable people.

The phrase 'Action and Reaction' makes it sound as though A acts on B and *then* B reacts on A. It is as though A is taking the initiative and B only strikes back in a kind of self-defense after it has been attacked – charge and countercharge, thrust and riposte, gambit and return.

This has led men into time-wasting blind alleys. They have reasoned: After A acts on B, it must take a finite (though possibly very small) time for B to react on A; and if I can only make the system do something after A acts on B but before B reacts on A then I can break the law of conservation of momentum or do something equally world-shaking.

The trouble is that there is no action and reaction with two bodies acting independently. The Third Law should be called 'The Law of Interaction' for both bodies act *together*.

Rather than argue it out, I will give you an analogous case. Suppose I tell you that Asimov's Law of Contact goes: If A touches B, then B touches A.

Do you think A first touches B and then B countertouches A? Do you think that there is a small but finite interval between A's touching of B and B's touching of A? Do you think that you can in any way relate the touch to either A or B alone? Or must you consider the touch as involving both together and inseparably?

Okay, then, I'm sure you get the point.

Now that we have the Laws of Motion, let's consider gravitation, which Newton also took up in *Principia Mathematica* and which was the subject of Chapter 5.

If, instead of throwing a ball, we held it in the air and simply let go, it would move downward with smoothly increasing speed. It would, in other words, accelerate downward. The Second Law of Motion holds that an acceleration can't exist without a force bringing it about. Therefore, to avoid breaking the Second Law, we find it necessary to postulate a 'gravitational force' in the direction of the Earth's center that acts on all masses. To symbolize the very special gravitational force, let us use the capital form of the letter and call it F.

If the strength of the gravitational force were fixed regardless of the nature of the falling body, then a more massive body would accelerate less (that is, would fall more slowly) than a less massive one. This would be expressed most simply in Equation 2.

But this is not so. The Italian scientist Galileo, nearly a century before the *Principia Mathematica* was published, conducted experiments which showed quite conclusively that all bodies, of whatever mass, accelerate equally as they fall (if we neglect air resistance).

Well, if Body A is twice as massive as Body B, then it takes twice the force to make Body A accelerate by a given amount as it would take to make Body B accelerate by that amount. If Body A is five times as massive as Body B, it takes five times the force for A as for B, and so on.

Therefore, if Galileo's demonstration is correct that all bodies of whatever mass accelerate downward equally as they fall, the gravitational force produced by the Earth is directly proportional to the mass of the falling body. Or:

$$F \sim m \qquad \text{(Equation 8)}$$

But by Newton's Third Law of Motion, if the Earth is exerting a force downward on the falling body, the falling body is exerting an equal force upward on the Earth.

This means that the Earth is accelerating upward as the falling body accelerates downward. However, the Earth is more massive than the falling body and accelerates to a correspondingly lesser degree. (I can hear you saying: But you just said that all bodies of whatever mass accelerate *equally*. Yes, to the Earth's gravitational pull. All bodies of whatever mass also accelerate equally to the falling body's gravitational pull, but the one 'equally' is not equal to the other 'equally.')

The Earth is so much more massive than the falling bodies we are accustomed to and therefore accelerates upward so much more slowly than the falling body accelerates downward that the Earth's acceleration goes unnoticed. This confuses the issue and it took good old Newton to cut through the complications and see that gravitation was a universal phenomenon and not a property of the Earth alone.

By the Third Law, if the Earth attracts the falling body, the falling body must attract the Earth in symmetrical fashion. If the attracting force depends on the mass of the falling body, it must also depend on the mass of the Earth, for we can't give one side special treatment over the other. If we let the mass of the Earth be represented as M, we can say:

$$F \sim mM \qquad \text{(Equation 9)}$$

The gravitational force also varies with the distance between the two bodies. It is reasonable to suppose that the farther apart the two bodies, the weaker their attraction for each other. We might argue, for instance, that the very simplest situation is that the gravitational force is inversely proportional to the distance of the falling body from the Earth. By the symmetry of the situation, though, it must also be inversely proportional to the distance of the Earth from the falling body. These two distances are obviously equal, and if one is represented by d so is the other. In that case, the gravitational force is inversely proportional to $d \times d$ or d^2; and is directly proportional to $1/d^2$. That is:

$$F \sim 1/d^2 \qquad \text{(Equation 10)}$$

And if we combine Equations 9 and 10, we have:

$$F \sim mM/d^2 \qquad \text{(Equation 11)}$$

To change the direct proportionality into an equality we want to multiply the right-hand side of the equation by a proportionality constant. In this case, let us call it the 'gravitational constant' and represent it as G. Equation 11 then becomes:

$$F = GmM/d^2 \qquad \text{(Equation 12)}$$

This represents Newton's law of gravitation, derived as simply as I could manage.

Let's see if we can simplify that equation. We can deal with the gravitational force in terms of acceleration. The Earth's acceleration is so inconceivably minute, we can ignore it and deal with the falling body's acceleration only. By Equation 6, we can substitute for F, the expression ma and then cancel m on both sides of the equation. This gives us:

$$a = GM/d^2 \qquad \text{(Equation 13)}$$

Can we get rid of G also? Well, Let's solve for G:

$$G = ad^2/M \qquad \text{(Equation 14)}$$

Unfortunately, that doesn't help us right away. We can measure the acceleration of the falling body (a) and we can measure the distance between the falling body and the center of the Earth (d), but we don't have the slightest idea as to the mass of the Earth (M). Or at least Newton didn't.

However, whatever G equals, its value remains the same for all possible values of a, d, and M, provided you always express a, d, and M in a fixed set of units. In that case, let's deliberately choose a convenient set of units that will enable us to get rid of G.

Suppose we use 'Earth mass' as the unit of mass and 'Earth radius' as the unit of distance and 'gravitational-acceleration unit' as the unit of acceleration. The Earth has a mass of exactly 1 Earth mass; the distance of the falling body from the center of the Earth is exactly 1 Earth radius; and the acceleration of the falling body is exactly 1 gravitational-acceleration unit. In that case:

$$G = 1 \times \frac{1^2}{1} = 1 \qquad \text{(Equation 15)}$$

As long as we keep those units, we can eliminate G and write Equation 13 as:

$$a = M/d^2 \qquad \text{(Equation 16)}$$

If we confine ourselves to the Earth, Equation 16 is utterly useless. All it tells us is that for a body with the size and the mass of the Earth, a falling body falls as fast as it is actually observed to fall. Big deal!

But what if we shift to the surface of the Moon. The mass of the Moon is 0.0124 times that of the Earth; that is, it is 0.0124 Earth masses. The distance of a falling object on the surface of the Moon from the Moon's center is equal to the Moon's radius which is 0.27 times that of the Earth's radius. We therefore find from Equation 16 (letting M now represent the mass of the Moon, and d the distance to the Moon's center):

$$a = 0.0124/0.27^2 = 0.17 \qquad \text{(Equation 17)}$$

We see that a falling body on the Moon's surface accelerates downward with 0.17 (roughly $\frac{1}{6}$) gravitational-acceleration units. Put more straightforwardly, a falling body accelerates $\frac{1}{6}$ as quickly on the Moon as on the Earth and, therefore (as one usually says), the surface gravity on the Moon is only $\frac{1}{6}$ that on the Earth.

And we know this without having to worry about the gravitational constant.

But we have gotten rid of the gravitational constant only by using very special units. We can actually convert gravitational-acceleration units into ordinary units involving centimeters and seconds by direct measurement. We can convert Earth radii into ordinary units involving centimeters, if we wish, by direct measurement. But what do we do with Earth masses?

That was a special problem which remained unsolved for a century after the *Principia Mathematica*. Then it was solved and, if you don't mind, I will end this chapter exactly as I ended the previous one:

The time was 1798, the place was England, the person was Henry Cavendish, the discussion thereof – well, please be patient.

7

The Man Who Massed the Earth

Just a few days ago I was at a dinner party and a nice lady, whom I did not know, cornered me and, for some reason unknown to myself, began telling me in superfluous detail of the manifold achievements of her son.

Now as it happens I have a very low attention span when the topic of conversation is something other than myself* and so I tried, rather desperately, to break the flow by asking some question or other.

The first that occurred to me was: 'And is this admirable young man your only son?'

To which the lady replied most earnestly, 'Oh, *no!* I also have a daughter.'

It had all been worth it, after all. The lady could not understand why I had broken into delighted laughter and even after I explained she had trouble seeing the humor of her reply.

Naturally, the juice of the situation was not just that the lady didn't hear me (that might have happened to anyone), but that it seemed to me to reflect, perfectly, the manner in which outmoded traditions of thought interfere with an understanding of the Universe as it is.

In pre-industrial society, for instance, male infants were much more valuable than female infants. Baby boys would grow into men and therefore presented, in potential, desperately needed help at the farm or in the army. Baby girls merely grew into women who had to be married off at great expense. Consequently, there was a great tendency to ignore daughters and to equate 'child' with 'son.'

The attitude still lingers, I think, even now, and even though

* I am told this, with varying degrees of mordacity (so look it up in the dictionary) by my nearest and dearest, but I maintain that this is not an evil peculiar to myself but is a common, and even necessary, attribute of writers generally.

the owner of such an attitude may be unaware of it and would deny its existence heatedly if accused of harboring it. I think that when the nice lady heard the phrase 'your only son' she honestly recognized no difference between that and 'your only child' and answered accordingly.

What has all this to do with this chapter? Well, scientists have similar problems and to this day they cannot free themselves utterly and entirely from outmoded ways of thought.

For instance, we all think we know what we mean when we speak of the 'weight' of something, and we all think we know what we mean when we say we are 'weighing' something or that one thing is 'heavier' or 'lighter' than another thing.

Except that I'm not at all sure we really do. Even physicists who are perfectly aware of what weight really is and can define and explain it adequately tend to slip into inaccurate ways of thought if not careful.

Let me explain.

The inevitable response to a gravitational field is an acceleration. Imagine, for instance, a material object suddenly appearing in space with no acceleration (relative to some large nearby astronomical body) at the moment of its appearance. Either it is motionless relative to that body or it is moving at a constant velocity.

If there were no gravitational field at the point in space where the body appeared, the body would continue to remain at rest or to move at constant velocity. If, however, there *is* a gravitational field at that point, as there must be from that large nearby astronomical body, the object begins to accelerate. It moves faster and faster, or slower and slower or it curves out of its original line of motion, or some combination of these.

Since in any universe that contains matter at all, a gravitational field (however weak) must exist at all points, accelerated motion is the norm for those objects in space which are subjected to gravitational fields only, and non-accelerated motion is an unrealizable ideal.

To be sure, if two objects are both accelerating precisely the same way relative to a third body, the two objects seem at rest with respect to each other. That is why you so often seem to yourself to be at rest. You *are* at rest with respect to Earth, but that is because both you and Earth are accelerating in response to the Sun's gravitational field in precisely the same way.

But then what about you and the *Earth*'s gravitational field. You may be at rest with respect to the Earth, but suppose a hole suddenly gaped below you. Instantly, in response to Earth's gravitational field, you would begin to accelerate downward.

The only reason you don't do so ordinarily is that there is matter solidly packed in the direction in which you would otherwise move and the electromagnetic forces set up by the atoms composing that matter hold those atoms together and easily block you from responding to the gravitational field.

In a sense, though, any material object prevented from responding to a gravitational field with an acceleration 'tries' to do so just the same.* It pushes in the direction it would 'like' to move in. It is this 'attempt' to accelerate in response to gravitation that makes itself evident as a force and it is this force which we can measure and call weight.

Suppose we use a coiled spring to measure force, for instance. If we pull at such a spring, the spring lengthens. If we pull twice as hard, it will lengthen twice as much. Within the limits of the spring's elasticity, the amount of lengthening will be proportional to the intensity of the force.

If, now, you fix one end of the spring to a beam in the ceiling and suspend a material object on the other end of the spring, the spring lengthens, just as though a force had been applied. A force *has* been applied. The material object 'tries' to accelerate downward and the force produced as a result of this 'attempt' lengthens the spring.

We can calibrate the spring by noting the amount of lengthening produced by bodies whose weights we have arbitrarily defined in terms of some standard weight somewhere. Once that is done, we can read off the weight of any object by having

* In this paragraph I am deliberately putting in quotes all the words that appear to give inanimate objects human desires and motivations. This is the 'bathetic fallacy' and it should be avoided except that it's such a convenient way of explaining things that sometimes I simply cannot resist being bathetic.

a pointer (attached to the lengthening spring) mark off a number on a scale.

All right, so far, but our notion of weight is derived, at its most primitve, from the feeling we have when an object rests on our hand or on some other part of our body and we must exert a muscular effort to keep it motionless with respect to Earth's gravitational field. Since we take Earth's gravitational field for granted and never experience any significant change in it, we attribute the sensation of weight entirely to the object.

An object is heavy, we think, because it is just naturally heavy and that's it, and we are so used to the thought that we don't allow ourselves to be disturbed by obvious evidence to the contrary. The weight of an object immersed in a liquid is decreased because the upward force of buoyancy must be subtracted from the downward force imposed by the gravitational field. If the buoyant force is great enough, the object will float and the denser the liquid the greater the buoyant force. Thus wood will float on water and iron will float on mercury.

We can actually feel an iron sphere to be lighter under water than in open air, yet we dismiss that. We don't think of weight as a force that can be countered by other forces. We insist on thinking of it as an intrinsic property of matter and when, under certain conditions, weight falls to zero, we are astonished, and we view the weightless cavortings of astronauts as something almost against nature. (They are 'beyond the reach of gravity' to quote the illiterate mouthings of too many newscasters.)

It is true that weight depends in part on a certain property innate in the object, but it also depends on the intensity of the gravitational field to which that object is responding. If we were standing on the surface of the Moon and were holding an object in our hand, that object would be 'attempting' to respond to a gravitational field that was only one sixth as intense as that on the surface of the Earth. It would therefore weigh only one sixth as much.

What is the innate property of matter on which weight partly depends? That is 'mass' (see Chapter 5), a term and concept which Newton introduced.

The force produced by a body 'attempting' to respond to a gravitational field is proportional to its mass as well as to the intensity of the gravitational field. If the gravitational field remains constant in intensity at all times (as is true, to all intents and purposes, of the Earth's gravitational field if we remain on or near its surface) we can ignore that field. We can then say that the force produced by a body 'attempting' to respond to Earth's gravitational field under ordinary circumstances is simply proportional to its mass.

(Actually, Earth's gravitational field varies from point to point, depending on the exact distance from the point to Earth's center and on the exact distribution of matter in the neighborhood of the point. These variations are far too tiny to detect through changes in the muscular effort required to counter the effect of weight, but they can be detected by delicate instruments.)

Since weight, under ordinary circumstances, is proportional to mass and vice versa, it is almost unbearably tempting to treat the two as identical. When the notion of mass was first established, it was given units ('pounds,' for instance) which had earlier been used for weight. To this day we speak of a mass of two kilograms and a weight of two kilograms and this is wrong. Units such as kilograms should be applied to mass only and weight should be given the units of force, but go talk to a brick wall.

The units have been so arranged that on Earth's surface, a mass of six kilograms also has a weight of six kilograms, but on the Moon's surface that same body will still have a mass of six kilograms and a weight of only one kilogram.

A satellite orbiting the Earth is in free fall with respect to the Earth and is already responding in full to Earth's gravitational field. There is nothing further for it to 'attempt' to do. Therefore a mass of six pounds on the satellite has a weight of zero pounds and the same is true of all objects, however massive. Objects on an orbiting satellite are therefore weightless. (To be sure, objects on an orbiting satellite ought to 'attempt' to respond to the gravitational fields of the satellite itself and of other objects on it, but these fields are so negligibly small, they can be ignored.)

Does it matter that the close match of weight and mass to which we are accustomed on the surface of Earth fails elsewhere? Sure it does. An object's inertia, that is the force required to accelerate it, depends entirely on its mass. A large metal beam is just as difficult to maneuver (to get moving when it is at rest, or to stop it when it is moving) on the Moon as on Earth, even though its weight is much less on the Moon. The difficulty of maneuver is the same on a space station even though weight is essentially zero.

Astronauts will have to be careful and if they don't forget Earth-born notions they may die. If you are caught between two rapidly moving beams you will be killed even though they are weightless. You will not be able to stop them with a flick of your finger even though they weigh less than a feather.

How can we measure mass? One way is to use the kind of balance consisting of two pans pivoting about a central fulcrum. Suppose an object of unknown weight is placed in the left pan. The left pan sinks and the right rises.

Suppose, next, a series of metal slivers, weighing exactly one gram each, are added to the right pan. As long as all the slivers, put together, weigh less than the unknown object, the right pan remains raised. When the sum of the slivers weighs more than the unknown, the right pan sinks and the left pan rises. When the two pans balance at the same level, the two weights are equal and you can say that the unknown weighs (let us say) seventy-two grams.

But now two weights at once are being subjected to the action of the gravitational field and the effect of that field cancels out. If the field is intensified or weakened, it is intensified or weakened on both pans simultaneously and the fact that the two pans are balanced is not affected. The two pans would remain in balance on the Moon, for instance. Such a balance is, therefore, to all intents and purposes measuring the one other property on which weight develops – mass.

Scientists prefer to measure mass rather than weight and so they train themselves to say 'more massive' and 'less massive' instead of 'heavier' and 'lighter' (though only with an effort and with frequent slips).

And yet they haven't freed themselves utterly from pre-Newtonian thinking even now, three centuries after Newton.

Picture this situation. A chemist carefully measures the mass of an object by using a delicate chemical balance and brings two pans into equilibrium as we have described. What has he done? He has 'measured the mass' of an object. Is there any shorter way of saying that correctly? No, there isn't. The English language doesn't offer anything. He can't say he has 'massed' the object, or 'massified' it or 'massicated' it.

The only thing he can say is that he has 'weighed' the object, and he *does* say it. I say it, too.

But to weigh an object is to determine its weight, not its mass. The unreformed English language forces us to be pre-Newtonian.

Again, these little slivers of metal that weigh a gram each (or any other convenient quantity or variety of quantities) should be called 'standard masses' if we are to indicate they are used in measuring mass. They are not. They are called 'weights.'

Again, chemists must frequently deal with the relative average masses of the atoms making up the different elements. These relative average masses are universally called 'atomic weights.' They are *not* weights, they are masses.

In short, no matter how well any scientist knows (in his head) the difference between mass and weight, he will never really know it (in his heart) as long as he uses a language in which hang-over traditions are retained. Like the lady who saw no difference between 'only son' and 'only child.'

Now let's move on. In the preceding chapters I talked about the masses of astronomical objects in terms of the mass of the Earth. Jupiter is 318 times as massive as Earth; the Sun is 330,000 times as massive as Earth; the Moon is 1/81 times as massive as Earth and so on.

But what is the mass of the Earth itself in kilograms (or any other unit of mass that we can equate with familiar everyday objects)?

To determine that we must make use of Newton's equation, presented in the previous chapter, which is:

$$F = GmM/d^2 \qquad \text{(Equation 1)}$$

If this equation is applied to a falling rock, for instance, F is the gravitational force to which the rock is responding by accelerating downward, G is the universal gravitational constant, m is the mass of the rock, M is the mass of the Earth, and d is the distance of the center of the rock from the center of the Earth.

Unfortunately, of the five quantities, the men of the eighteenth century could only determine three. The mass of the rock (m) could easily be determined, and the distance of the rock from the center of the Earth (d) was known as far back as the time of the ancient Greeks. The gravitational force (F) could be determined by measuring the acceleration with which the rock was responding to the gravitational field and that had been done by Galileo.

Only the values of G, the gravitational constant, and M, the mass of the Earth, remained unknown. If only the value of G were known, the mass of the Earth could be calculated at once. Conversely, if M were known, the universal gravitational constant could be quickly determined.

What to do?

The mass of the Earth could be determined directly if it could be manipulated; if it could be placed on a balance pan against standard weights or something like that. However, the Earth cannot be manipulated, at least by men in a laboratory, so forget that.

Then what about determining G. This is the universal gravitational constant and it is the same for *any* gravitational field. That means we don't have to use the Earth's gravitational field to determine it. We might instead use the gravitational field of some smaller object which we can freely manipulate.

Suppose, for instance, we suspend an object from a spring and lengthen the spring thanks to the effect of Earth's gravitational field. Next, we take a large boulder and place it under the suspended object. The gravitational field of the boulder is now added to the Earth's gravitational field and the spring is extended a little further as a result.

From the amount of the additional lengthening of the spring, we could determine the intensity of the gravitational field of the boulder.

Now let us use the following variation of Newton's equation:

$$f = Gmm'/d^2 \qquad \text{(Equation 2)}$$

where f is the gravitational field intensity of the boulder (measured by the additional extension of the spring), G is the gravitational constant, m the mass of the object suspended from the spring, m' the mass of the boulder, and d the distance between the center of the boulder and the center of the suspended object.

Every one of these quantities can be determined except G, so we rearrange Equation 2 thus:

$$G = fd^2/mm' \qquad \text{(Equation 3)}$$

and at once have the value of G. Once we know that value we can substitute it in Equation 1, which we can then solve for M (the mass of the Earth), as follows:

$$M = Fd^2/Gm \qquad \text{(Equation 4)}$$

But there is a catch. Gravitational fields are so incredibly weak in relation to mass that it takes a hugely massive object to have a gravitational field intense enough to measure easily. The boulder held under the suspended object would simply not produce a measurable further extension of the spring, that's all.

There is no way of making the gravitational field more intense, so if the problem of the mass of the Earth was to be solved at all, some exceedingly delicate device would have to be used. What was needed was something that would measure the vanishingly small force produced by the vanishingly small gravitational field produced by an object small enough to be handled in the laboratory.

The necessary refinement in measurement came about with the invention of the 'torsion balance' by the French physicist Charles Augustin Coulomb in 1777 and (independently) by the English geologist John Michell, as well.

Instead of having a force extend a spring or twist a pan about a fulcrum, it was used to twist a string or wire.

If the string or wire was very fine, only a tiny force would be required to twist it quite a bit. To detect the twist, one need attach to the vertical wire a long horizontal rod balanced at the center. Even a tiny twist would produce a large movement at the end of the rods. If a thin wire is used and a long rod, a torsion balance could be made enormously delicate, delicate enough even to detect the tiny gravitational field of an ordinary object.

In 1798, the English chemist Henry Cavendish put the principle of the torsion balance to use in determining the value of G.

Suppose you take a rod six feet long and place on each end a two-inch-in-diameter lead ball. Suppose you next suspend the rod from its center by a fine wire.

If a very small force is applied to the one lead ball on one side and an equally small force to the other lead ball on the other side, the horizontal rod will rotate and the wire to which it is attached will twist. The twisting wire 'attempts' to untwist. The more it is twisted the stronger the force to untwist becomes. Eventually, the force to untwist balances the force causing it to twist and the rod remains in a new equilibrium position. From the extent to which the rod's position has shifted, the amount of force upon the lead balls can be determined.

(Naturally, you must enclose the whole thing in a box and place it in a sealed, constant-temperature room so that no air currents – produced either by temperature differences or mechanical motions – confuse the situation.)

Where the rod takes up only a slightly different position it means that even a tiny twist of the fine wire produces enough counterforce to balance the applied force. What a tiny force it must then be that was applied – and that was exactly what Cavendish had in mind.

He suspended a lead ball eight inches in diameter on one side of one of the small lead balls at the end of the horizontal rod. He suspended another such ball on the opposite side of the other small lead ball.

The gravitational field of the large lead balls would now serve to twist the rod and force it into a new position (see Figure 1).

Cavendish repeated the experiment over and over again and from the shift in the position of the rod and, therefore, from the twist of the wire, determined the value of f in Equation 3.

Since he knew the values of m, m' and d, he could calculate the value of G at once.

Cavendish's value was off by less than 1 per cent from the value now accepted, which is 0.0000000000667 meters^3/kilogram-second^2. (Don't ask about the significance of that unit; it is necessary to make the equations balance.)

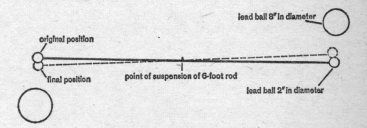

Once we have the value for G in the units given, we can solve Equation 4, and if we use the proper units out will pop the mass of the Earth in kilograms. This turns out to be 5,983,000,000,000,000,000,000,000 or 5.983×10^{24} kilograms. (If you want it, roughly, in words, say, 'about six septillion kilograms.')

Once we have the Earth's mass in kilograms, we can determine the mass of other objects, too, provided only their mass relative to that of Earth is known.

The Moon, which has a mass 1/81 that of the Earth, has a mass of 7.4×10^{22} kilograms. Jupiter, with a mass 318 times that of Earth, has a mass of 1.9×10^{27} kilograms. The Sun, with a mass 330,000 times that of Earth, has a mass of 2×10^{30} kilograms.

Thus, Cavendish not only measured the mass of the Earth, but he measured (at least potentially) the mass of every other object in the Universe just by noticing the small shift in position of a pair of lead balls when a pair of larger lead balls was placed nearby.

How's that for the power of a simple equation?

But – and there is the point of the whole chapter – when someone wishes to mention this astonishing achievement of Cavendish's, what does he say? He says: 'Cavendish weighed the Earth.'

Even physicists and astronomers speak of Cavendish as the man who 'weighed the Earth.'

He did no such thing! He determined the *mass* of the Earth. He *massed* the Earth. It may be that English has no such verb, but that's the fault of the language, not of me. To me, Cavendish is the man who massed the Earth and English can like it or lump it.

Which leaves one question: What *is* the weight of the Earth?

The answer is simple. The Earth is in free fall and, like any object in free fall, it is responding in full to the gravitational fields to which it is subject. It is not 'attempting' to make any further response and therefore it is weightless.

The weight of the Earth, then, is zero.

8

The Luxon Wall

You wouldn't think my science essays would get a mention in *Time*, would you?* Well, they did, and the particular article that was mentioned, some months ago, was 'Impossible, That's All' (see Chapter 4 of *Science, Numbers and I*, Doubleday & Company, 1968).

That article dealt with the impossibility of attaining or surpassing the velocity of light. After the article was published, there came to be a great deal of talk about faster-than-light particles and suddenly I sounded like a fuddy-duddy left flat-footed by the advance of physics past the bounds I had mistakenly thought fixed.

At least that's the way *Time* made me sound. To make it even worse, they cited my good old friend Arthur C. Clarke,† and his rebuttal entitled 'Possible, That's All' in a way that sounded as though they thought Arthur more forward-looking than myself.

Fortunately, I am a tolerant man who is not disturbed by such things and I shrugged it off. When I next met Arthur, we were still the best of friends if you don't count the kick in the shin I gave him.

Anyway, I am *not* a fuddy-duddy and I am now going to explain the situation in greater detail in order to prove it.

Let's begin with an equation that was first worked out by the Dutch physicist Hendrik Antoon Lorentz in the 1890s. Lorentz thought it applied specifically to electrically charged bodies but Einstein later incorporated it into his Special Theory of Relativity, showing that it applied to all bodies, whether they carried an electric charge or not.

* Come to think of it, why not?
† He's three years older than I am. I thought I'd just casually mention that.

I will not present the Lorentz equation in its usual form but will make a small change for purposes that will eventually become clear. My version of the equation, then, is as follows:

$$m = k/\sqrt{1 - (v/c)^2} \qquad \text{(Equation 1)}$$

In Equation 1, m represents the mass of the body under discussion, v is the velocity with which it is moving with respect to the observer, c is the velocity of light in a vacuum, and k is some value that is constant for the body in question.

Suppose, next, that the body is moving at one-tenth the velocity of light. That means $v = 0.1c$. In that case, the denominator of the fraction on the right-hand side of Equation 1 becomes $\sqrt{1 - (0.1c/c^2} = \sqrt{1 - 0.1^2} = \sqrt{1 - 0.01} = \sqrt{0.99} = 0.995$. Equation 1 therefore becomes $m = k/0.995 = 1.005k$.

We can make the same sort of evaluation for the case of the same body at gradually increasing velocities, say at velocities equal to $0.2c$, $0.3c$, $0.4c$ and so on. I won't bore you with the calculations but the results come out as follows:

Velocity	Mass
0.1c	1.005k
0.2c	1.03k
0.3c	1.05k
0.4c	1.09k
0.5c	1.15k
0.6c	1.24k
0.7c	1.41k
0.8c	1.67k
0.9c	2.29k

As you see, the Lorentz equation, if correct, would indicate that the mass of any object increases steadily (and, indeed, more and more rapidly) as its velocity increases. When this was first suggested, it seemed utterly against common sense because no such change in mass had ever been detected.

The reason for the non-detection, however, lay in the fact that the value of c was so huge by ordinary standards – 186,281 miles per second. At a velocity of only one tenth the speed of light, the mass of an object has increased to one half of one per cent more than its mass at, say, sixty miles per hour, and this increase would be easily detectable in principle. However, a velocity of 'only' one tenth the speed of light (0.1c) is still 18,628 miles per second or over 67 million miles per hour. In other words, to get measurable changes in mass, velocities must be attained which were completely outside the range of experience of the scientists of the 1890s.

A few years later, however, subatomic particles were detected speeding out of radioactive atomic nuclei and their velocities were sometimes considerable fractions of the velocity of light. Their masses could be measured quite accurately at different velocities and the Lorentz equation was found to hold with great precision. In fact, down to this moment, no violation of the Lorentz equation has ever been discovered for any body at any measured velocity.

We must, therefore, accept the Lorentz equation as a true representation of that facet of the Universe which it describes – at least until further notice.

Now, accepting the Lorentz equation, let's ask ourselves some questions. First, what does k represent?

To answer that, let's consider a body (any body that possesses mass) that is at rest relative to the observer. In that case, its velocity is zero and since $v = 0$, then $v/c = 0$ and $(v/c)^2 = 0$. What's more, $\sqrt{1 - (v/c)^2}$ is then $\sqrt{1 - 0}$ or $\sqrt{1}$ or 1.

This means that for a body at rest relative to the observer, the Lorentz equation becomes $m = k/1 = k$. We conclude then that k represents the mass of a body that is at rest relative to the observer. This is usually called the 'rest-mass' and is symbolized as m_0. To write the Lorentz equation in the form it is usually seen, then, we have:

$$m = m_0/\sqrt{1 - (v/c)^2} \qquad \text{(Equation 2)}$$

The next question is what happens if an object moves at velocities higher than the highest velocity given in the small table presented earlier in the article. Suppose the object moved at a velocity of 1.0c relative to an observer; in other words, if it moved at the velocity of light.

In that case the denominator of the Lorentz equation becomes $\sqrt{1 - (1.0c)^2} = \sqrt{1 - 1^2} = \sqrt{1 - 1} = \sqrt{0} = 0$. For a body moving at the speed of light, the Lorentz equation become $m = m_0/0$ and if there is one thing we are not allowed to do in mathematics, it is to divide by zero. Mathematically, the Lorentz equation becomes meaningless for a body possessing mass that is moving at the velocity of light.

Well, then, let's sneak up on the velocity of light and not try to land right on it with a bang.

As one increases the value of v, in Equation 2, past $0.9c$, while keeping it always *less* than $1.0c$, the value of the denominator steadily approaches zero and, as it does so, the value of m gets larger without limit. This is true no matter what the value of m_0 so long as that value is greater than zero. (Try it for yourself, calculating m for values of v equal to $0.99c$, $0.999c$, $0.9999c$ and so on for as long as you have patience.)

In mathematical language, we would say that in any fraction $c = a/b$, where a is greater than 0, then, as b approaches zero, c increases without limit. A shorthand way of saying this, and one that is frowned upon by strict mathematicians, is $a/0 = \infty$, where ∞ represents increase without limit or 'infinity.'

So we can say, then, that for any object possessing mass (however little) that mass approaches infinite values as its velocity approaches the velocity of light relative to the observer.

This means that the body cannot actually attain the velocity of light (though it can fall only infinitesimally short of that value) and certainly cannot surpass it. You can show this by either of two lines of argument.

The only way we know of by which an ordinary mass-possessing object can be made to go at a greater velocity than it already possesses is to apply a force and therefore produce an acceleration (see Chapter 6). The greater the mass, however, the smaller the acceleration produced by a given force and, therefore, as the mass approaches infinite values the acceleration which it can attain when subjected to any force, however great, approaches zero. Consequently, the object cannot be made to go faster than the velocity at which its mass becomes infinite.

The second line of argument is as follows. A moving body possesses kinetic energy which we may set equal to $mv^2/2$, where m is its mass and v is its velocity. If a force is applied to the body so that its kinetic energy is increased, that energy may

100

be increased because v is increased, or because m is increased, or because both v and m are increased. At ordinary, everyday velocities, all the measurable increase goes into velocity, so we assume (wrongly) that mass remains constant under all conditions.

Actually, though, both velocity and mass are increased as the result of an applied force, but the mass so slightly at ordinary velocities as to make the change immeasurable. As the velocity relative to the observer increases, however, a larger and larger fraction of the added energy produced by an applied force goes into increasing the mass and a smaller and smaller fraction into increasing the velocity. By the time the velocity is very close to that of light, virtually all the energy increase appears in the form of an increase in mass; virtually none in the form of an increase in velocity. The change in emphasis is such that the final velocity can never attain, let alone exceed, the velocity of light.

And don't ask why. That's the way the Universe is constructed.

I hope you notice, though, that when I was discussing the fact that mass becomes infinite at the speed of light I was forced by the mathematical facts of life to say, 'This is true no matter what the value of m_0 so long as that value is greater than zero.'

Of course, all the particles that build up ourselves and our devices – protons, electrons, neutrons, mesons, hyperons, etc. etc. – have rest masses greater than zero so that the restriction doesn't seem very restrictive. In fact, people generally say, 'It is impossible to attain or surpass the velocity of light,' without specifying that they mean for objects possessing rest mass greater than zero because that seems to include virtually everything anyway.

I neglected to make the restriction myself in 'Impossible, That's All,' which is what left me open to the fuddy-duddy implication. If we *include* the restriction, then everything I said in the article is perfectly valid.

Now let's go on and consider bodies with m_0 not greater than zero.

Consider a photon, for instance, a 'particle' of electromagnetic radiation – visible light, microwaves, gamma rays, etc.

What do we know about photons? In the first place, a photon always possesses finite energy so that its energy content is somewhere between 0 and ∞. Energy, as Einstein showed, is equivalent to mass according to a relationship which he expressed as $e = mc^2$. This means that any photon can be assigned a mass value which can be calculated by this equation and which will also fall somewhere between 0 and ∞.

Another thing we know about photons is that they move (relative to any observer) at the velocity of light. Indeed, light has that velocity because it is made up of photons.

Now that we know these two things, let us convert Equation 2 into another but equivalent form:

$$m\sqrt{1 - (v/c)^2} = m_0 \qquad \text{(Equation 3)}$$

For a photon, $v = c$, and by now you should see at once that this means that, for a photon, Equation 3 becomes:

$$m(0) = m_0 \qquad \text{(Equation 4)}$$

If a photon were an ordinary mass-possessing object and were traveling at the velocity of light, its mass (m) would be infinite. Equation 4 would then become $\infty \times 0 = m_0$ and such an equation is not permitted in mathematics.

A photon, however, can be assigned a value for m which is between 0 and ∞, even though it is traveling at the velocity of light, and for *any* value between 0 and ∞ assigned to m, the value of m_0 in Equation 4 turns out to equal 0.

This means that for photons, then, the rest mass (m_0) equals zero. If the rest mass is zero, in other words, an object *can* move at the velocity of light.

(This should dispose of the perennial question I am asked by correspondents who think they have discovered a flaw in Einstein's logic and say: 'If anything moving at the velocity of light has infinite mass, how come photons don't have infinite mass?' The answer is that a distinction must be made between particles with a rest mass of 0 and particles with a rest mass greater than 0. But don't worry. Correspondents will continue to ask the question no matter how often I explain.)

But let's go farther. Suppose a photon traveled at a velocity *less* than that of light. In that case the quantity under the square root sign in Equation 3 would be greater than zero and this would be multiplied by m which itself possesses a value greater than zero. If two values, each greater than zero, are multiplied, then the product (in this case, m_0) must be greater than zero.

This means that if a photon traveled at less than the velocity of light (no matter how infinitesimally less), it would no longer have a rest mass equal to zero. The same would be true if it traveled at more than the velocity of light, no matter how infinitesimally more. (Funny things happen to the equation at velocities greater than light as we shall soon see, but one thing that the funniness cannot obscure is that the rest mass would no longer be equal to zero.)

Physicists insist that a rest mass must be constant for any given body, since all the phenomena they measure make sense only if this is so. In order for a photon's rest mass to remain constant, then (that is, for it to remain always zero) the photon must *always* move at the velocity of light, not a hair less and not a hair more provided it is moving through a vacuum.

When a photon is formed, it *instantly*, with no measurable time lapse, begins moving away from the point of origin at 186,281 miles per second. This may sound paradoxical because it implies an infinite rate of acceleration and therefore an infinite force but stop —

Newton's second law, connecting force, mass and acceleration, applies only to bodies with a rest mass greater than zero. It does *not* apply to bodies with a rest mass equal to zero.

Thus, if energy is poured into an ordinary body under ordinary circumstances, its velocity increases; if energy is subtracted, its velocity decreases. If energy is poured into a photon, its frequency (and mass) increases but its velocity remains unchanged; if energy is subtracted, its frequency (and mass) decreases but its velocity remains unchanged.

But if all this is so, it seems to make poor logic to speak of 'rest mass' in connection with photons, for that implies the mass a photon would have if it were at rest – and a photon never can be at rest.

An alternate term has been suggested by O.-M. Bilaniuk and E. C. G. Sudarshan.* This is 'proper mass.' The proper mass of an object would be a constant mass value that is inherent in the body and is not dependent on velocity. In the case of ordinary bodies, this inherent mass is equal to that which can be measured when the body is at rest. In the case of photons, it can be worked out by deduction, rather than by direct measurement.

A photon is not the only body that can and *must* travel at the velocity of light. Any body with a proper mass of zero can and must do so. In addition to photons, there are no less than five different kinds of particles that are thought to have a proper mass of zero.

One of these is the hypothetical graviton, which carries the gravitational force and which, in 1969, may have finally been detected.

The other four are the various neutrinos: (1) the neutrino itself, (2) the antineutrino, (3) the muon-neutrino, and (4) the muon-antineutrino.

The graviton and all the neutrinos can and must travel at the velocity of light. Bilaniuk and Sudarshan suggest that all these velocity-of-light particles be lumped together as 'luxons' (from the Latin word for 'light').

All particles with a proper mass greater than zero, which therefore cannot attain the velocity of light and must always and forever travel at lesser velocities, they lump together as 'tardyons.' They further suggest that tardyons be said to travel always at 'subluminal' ('slower-than-light') velocities.

But now what if we think of the unthinkable and consider particles at 'superluminal' ('faster-than-light') velocities? This was done for the first time with strict adherence to relativistic principles (as opposed to mere science-fictional speculation) by

* In an article entitled 'Particles Beyond the Light Barrier,' *Physics Today*, May 1969, for those of you who wish I would give a reference now and then.

Bilaniuk, Deshpande and Sudarshan in 1962, and such work hit the headlines at last when Gerald Feinberg published a similar discussion in 1967. (It was Feinberg's work which inspired the discussion in *Time*.)

Suppose a particle traveled at a velocity of $2c$; that is, at twice the velocity of light. In that case, v/c would become $2c/c$ or 2 and $(v/c)^2$ would be 4. The term $\sqrt{1 - (v/c)^2}$ would become $\sqrt{1-4}$ or $\sqrt{-3}$ or $\sqrt{3}\sqrt{-1}$.

Since it is usual to express $\sqrt{-1}$ as i and since $\sqrt{3}$ is approximately 1.73, we can say that for a particle traveling at the velocity twice that of light, Equation 3 becomes:

$$1.73mi = m_0 \qquad \text{(Equation 5)}$$

Any expression that contains i (that is $\sqrt{-1}$) is said to be imaginary, a poor name but one that is ineradicable.

It turns out, as you can see for yourself if you try a few random examples that for any object traveling at superluminal velocities, the proper mass is imaginary.

An imaginary mass has no physical significance in our own subluminal Universe, so it has long been customary to dismiss superluminal velocities at once, and say that faster-than-light particles are impossible because there can be no such thing as an imaginary mass. I've said it myself in my time.

But is an imaginary mass truly without meaning? Or is a mass represented by mi merely a mathematical way of expressing a set of rules not like the rules we are accustomed to, but rules that *still* obey the dictates of Einstein's special relativity?

Thus, in the case of games such as baseball, football, basketball, soccer, hockey and so on and so on, the contestant or contestants who make the higher score win. Yet can one say from this that a game in which the lower score wins is unthinkable? How about golf? The essential point in any game of skill is that the contestant who achieves the more difficult task wins – the more difficult task usually involving a higher score, but in golf involving a lower score.

In the same way, in order to obey special relativity, an object with an imaginary rest mass must behave in ways which seem paradoxical to those of us who are used to the behavior of objects with real rest masses.

For instance, it can be shown that if an object with an imaginary rest mass increases in energy, its velocity *decreases*; if it decreases in energy, its velocity *increases*. In other words, an object with an imaginary rest mass will slow down when a force is applied and speed up when resistance is encountered.

Furthermore, as such particles add energy and slow down, they can never quite slow down to the velocity of light. At the velocity of light their mass becomes infinite. As their energy decreases to zero, however, their velocity increases without limit. A body with imaginary rest mass, which possesses zero energy, would have infinite velocity. Such particles move always faster than light and Feinberg has suggested they be termed 'tachyons' from a Greek word meaning 'fast.'

Well, then, the tardyon Universe is subluminal, with possible velocities ranging from 0 for zero energy to c for infinite energy. The tachyon Universe is superluminal, with possible velocities ranging from c for infinite energy to ∞ for zero energy. Between the two Universes is the luxon Universe, with possible velocities confined to c, and never either less or more, at any energy.

We might view the total Universe as divided into two compartments by an unbreachable wall. We have the tardyon Universe on one side, the tachyon Universe on the other side, and between them, the infinitely thin but infinitely rigid luxon wall.

In the tardyon Universe most objects have little kinetic energy. Those objects that have great velocities (like a cosmic-ray particle) have very little mass. Those objects with great masses (like a star) have very little velocity.

The same is very likely true in the tachyon Universe. Objects with relatively slow velocities (just slightly more than light) and therefore great energies must have very little mass and be not too different from our cosmic-ray particles. Objects with great masses would have very little kinetic energy and therefore enormous velocities. A tachyon star might be moving at trillions of times the velocity of light, for instance. But that

106

would mean that the mass of the star would be distributed over vast distances through tiny time intervals so that very little of it would be present in any one place at any one time, so to speak.

The two universes can impinge and become detectable, one to the other, at only one place – the luxon wall at which they meet. (Both universes hold photons, neutrinos and gravitons in common.)

If a tachyon is energetic enough and therefore moving slowly enough, it might have sufficient energy and hang around long enough to give off a detectable burst of photons. Scientists are watching for those bursts, but the chance of happening to have an instrument in just the precise place where one of those (possibly very infrequent) bursts appears for a billionth of a second or less is not very great.

Of course, we might wonder whether there might not be some possibility of breaching the luxon wall by some means less direct than accelerating past it – which is impossible (that's all). Can one turn tardyons into tachyons somehow (possibly by way of photons) so that suddenly one finds one's self transferred from one side of the wall to the other without ever having gone through it? (Just as one can combine tardyons to produce photons and suddenly have objects moving at the velocity of light without having been accelerated to it.)

The conversion to tachyons would be equivalent to the entry into 'hyperspace,' a concept dear to science fiction writers. Once in the tachyon Universe, a spaceship with the energy necessary to go only a tiny fraction of the velocity of light would find itself going (with the same energy) at very many times the velocity of light. It could get to a distant galaxy in, say, three seconds, then turn back automatically to tardyons and be in our own Universe again. That would be equivalent to the interstellar 'jump' that I am always talking about in my own novels.

In connection with that, though, I have an idea which, as far as I know, is completely original with me. It is not based on any consideration of physical law but is purely intuitive and arises only because I am convinced that the overriding characteristic of the Universe is its symmetry and that its overriding principle is the horrid doctrine of 'You can't win!'

I think that each Universe sees itself as the tardyon Universe and the other as the tachyon Universe, so that to an observer from neither (perched on top of the luxon wall, so to speak) it would appear that the luxon wall separates identical twins, really.

If we managed to transfer a spaceship into the tachyon Universe, we would find ourselves (I intuitively feel) still going at subluminal velocities by our new measurements and looking back at the Universe we had just left as superluminal.

And if so, then whatever we do, *whatever* we do, tachyons and all, attaining or surpassing the velocity of light will remain impossible, that's all.

9

Playing the Game

My occupation as a writer on far-out subjects makes it inevitable that Gentle Readers send me all kinds of far-out letters, pamphlets and books. Their material is welcome and I read everything sent me up to the point where I judge it to be no longer worth reading. Sometimes this means I read the whole thing and profit and sometimes I only read a few paragraphs.

It is not often, however, that I find my mental indicator veering over to 'worthless' before I even start the book and indeed at the very first paragraph of the publisher's blurb. Exactly that happened recently.

I had received a book purporting (according to the covering letter) to reveal the truth about the origins of the Universe and to expose such fakers as Newton and Einstein. I must admit the letter itself left me without too much hope. However, fair's fair – I opened the book-cover and glanced at what was printed on the inner flap of the book jacket.

It listed some of the startling points made by the author and the very first was that the author 'believes' that light which travels long distances is gradually absorbed; that the blue end of the spectrum is absorbed first, and that this accounts for the 'red shift' of the light from distant galaxies.

Now if this belief were true, then all astronomical deductions based on the red shift must be re-evaluated and there would be a chance that our most basic opinions concerning the Universe would be completely overturned. Surely I ought to find such beliefs exciting in the extreme and should therefore read on.

But no, for having read that much of the book flap I put the book itself aside for disposal. What else could I do? Under certain conditions, light traveling long distances *is* absorbed or scattered, and blue light *is* absorbed or scattered first – but that has nothing to do with the red shift. Since the author clearly didn't know what the red shift was, what good were his beliefs on the subject?

What good are beliefs by themselves anyway? There is a view very popular among amateurs in science that it is only necessary to have a 'theory' in order to revolutionize science. In actual fact, 'theories' by themselves are nothing but intellectual amusement and to become more than that they must be supported by observations; preferably by observations that not only support the theory but that quash opposing theories.

Science, like other intellectual games, has its rules – rules that have been strictly applied for nearly four centuries – and the results attained are ample evidence that the rules are good ones. Those who would revolutionize science had better learn the rules; not because that will make them respectable but because, believe me, they will never revolutionize science without them.

It is odd that, though no one who has never studied chess would dream he could beat a Grand Master, so many strict amateurs with little or no scientific training are convinced they can point out the 'obvious' flaws in Einstein's theories.

Nor can amateurs console themselves (as they often do) with the thought that 'they laughed at Galileo.' Sure, some did, but *many didn't*. Galileo overthrew Aristotelian physics because, for one thing, he was a thoroughgoing student of that same Aristotelian physics. Similarly, Copernicus upset the Ptolemaic theory because, in part, he had a thorough education in Ptolemaic theory; and Vesalius cast out Galenic anatomy because, it must be understood, he was an expert on Galenic anatomy.

This is a general rule that must be understood by revolutionaries (perhaps in all fields, but certainly in science). You must thoroughly know that which you hope to supplant.

And now, as an example of how the game *is* played, let us go on to this matter of the red shift and show just how that was worked out, how long it took, how many contributed, and how unlikely it is to be pounded into new shape by the vague speculations of the untutored.

The matter begins with a phenomenon that may have gone largely unnoticed until the coming of the railroad.

Suppose a whistle is emitting a sound of a given pitch. If the whistle is approaching you as it is sounding, the pitch it produces is higher than it would be if the whistle were motionless with respect to you. If, on the other hand, the whistle were moving away from you, the pitch would be lower than it would be if the whistle were motionless. What's more, the difference in pitch would be greater the more rapidly the whistle was moving toward you or away from you.

So far, I am just making flat statements. Are they true?

Well, before the 1830s nobody was likely to have wondered whether such statements were true or not. Probably nobody ever noticed any such pitch changes.

Prior to the coming of the railroad the most rapid motion one was likely to encounter was that of a galloping horse. We might imagine a man on horseback blowing a sustained note on a horn. If the observer watched such a man ride past, the note would be higher than it would ordinarily be while he was approaching, lower than it would ordinarily be while he was departing. There would be a sudden drop when he was passing.

However, galloping horses with horn-blowing riders weren't all that common and where such a situation did exist, the only explanation I can think of is that the horseman was probably leading a charge in battle. With hundreds of spear-wielding or saber-swinging cavalrymen galloping after him, no potential observer is going to be listening to the pitch of the horn.

But then came the locomotive, with its steam whistle designed to perform the amiable work of warning men and beasts to get off the track! The locomotive passed periodically; its intentions were peaceful; observers watched and listened in fascination. They heard the steam whistle switch from tenor to baritone as it passed.

After that happened a number of times, curiosity was bound to be aroused. Why did it happen? Why did the whistle sound shriller when the train was approaching than when the train was standing at the station? Why did the whistle sound deeper when the train was departing? Why the sudden change just as it passed?

One could eliminate at once the thought that the engineer was playing tricks, since it happened with all trains and all

111

engineers and such elaborate trickery for no reason is utterly implausible. Besides, if two people were listening from different vantage points, the whistle was still shrill to those people it hadn't reached when it had already dropped in pitch to those people it had passed.

Enter an Austrian physicist, Christian Johann Doppler, who, in 1842, undertook to explain the phenomenon in terms of the nature of sound itself.

Sound, it was by then understood, was a wave phenomenon, producing its effect on the human ear through the existence of alternate regions of compression and rarefaction traveling across the medium through which it was transmitted. The distance from one region of compression to the next is called the 'wavelength' and it could be shown that the shorter the wavelength, the higher the pitch, and the longer the wavelength, the lower the pitch.

It so happens that sound, at 0° C., travels at a velocity of close to 330 meters per second.* This means that each region of compression is moving outward from the sound source, in all directions, at 330 meters per second.

Suppose, now, that a whistle is emitting a steady blast at a given pitch, one in which 330 waves of compression are produced each second – a pitch equivalent to E above middle C on the piano. One second after the note has begun sounding, 330 waves of compression have been formed. The first of these waves has had the opportunity to move outward 330 meters and the rest are all strung out evenly behind it. You can see that this means there is exactly 1 meter distance between the successive compressions and that the wavelength of the sound is just 1 meter.

But suppose the whistle is on a locomotive that is approaching you at thirty-three meters per second, which is one tenth the speed of sound.

By the time the first wave of compression has traveled 1

* I am going to use 'meters per second' as the unit of velocity throughout this chapter just to be devilish (and scientifically conventional). If you are more comfortable with 'miles per hour,' just keep in mind that 1 meter per second equals just about 2¼ miles per hour. Thus, 330 meters per second equals 752 miles per hour.

meter from its starting point, the second wave of compression is ready to be emitted, and by that time the locomotive has traveled forward 0.1 meters. The second wave of compression starts out, therefore, only 0.9 meters behind its predecessor, and this same argument applies for every wave of compression that follows.

In other words, where a particular locomotive whistle emits sound with a wavelength of 1 meter when it is at rest with respect to you, it emits sound with a wavelength of 0.9 meters when it is approaching you at a speed of 33 meters per second. (And you can see for yourself that the wavelength would be shorter still if it were approaching still more rapidly.)

A wavelength of 0.9 meters is equivalent to a frequency of 330/0.9 or 367 per second, which is almost the equivalent of F-sharp.

Everything is just the reverse if the locomotive is moving away from you. Then when the first wave of compression has traveled 1 meter toward you and the second wave is about to be emitted, the locomotive has traveled 0.1 meter away from you and the wavelength is 1.1 meters. The frequency is 330/1.1 or 300 per second, which is just about equivalent to the note D.

We can say, then, that a locomotive whose whistle naturally sounds E, sounds F-sharp as it approaches at the usual speed of a moderately fast express and that the sound drops suddenly to D as it passes.

Notice that the pitch is related to the relative motion of locomotive and observer. That may seem surprising. After all, if the locomotive is approaching and is squeezing the compression waves closer together what has that to do with the person listening? Well, if you are moving in the same direction as the locomotive and at the same speed, the waves of compression, which are pushed together as they emerge from the whistle, would be stretched apart again as they race to overtake your ears. And if you and the train are moving at the same velocity, the stretching at the ear would exactly balance the compression at the whistle. You would hear the normal pitch exactly as though both you and whistle were motionless.

So far, though, Doppler's analysis (which was essentially the above) is pure ivory-tower theorizing. Still, it offers a possibility for observation. Doppler's theory related the speed of sound, the speed of the sound source and the pitch of the sound in a perfectly definite way, and it was only necessary to run an experiment that would allow the necessary measurements to be made in order to see if they were indeed in accord with the Doppler analysis.

To do that, Doppler managed to commandeer a flatcar and a stretch of Dutch railway in 1844. On the flatcar, he placed trumpeters with orders to sound this note or that and to hold it steady as they passed observers with a sense of absolute pitch. The observers reported the exact drop in pitch and from the speed of sound and of the flatcar Doppler did some calculating and found that the observed drop in pitch was equal to that predicted by his theory. This change of pitch with velocity has been called the 'Doppler effect' ever since.

Doppler saw at once that his analysis could apply not to sound alone, but to any wave form at all. Light, for instance, consisted of tiny waves, and if a light source emitting a single wavelength of light were approaching you, surely that wavelength should shorten and the frequency increase. If the source receded from you, the wavelength should lengthen and the frequency decrease.

Whereas sound alters in pitch as the wavelength changes, light alters in color. Visible light can have any of a range of wavelengths, the longest representing the color red, and then, with decreasing wavelength, orange, yellow, green, blue and violet (the spectrum, with the rainbow as a natural example).

Well, then, if a receding light source emits a particular wavelength, that wavelength is longer than it would be if the light source were stationary. Its color would shift in the direction of the red. If the light source were approaching, its color would shift in the direction of the violet.

We can say, then, that a receding light source is accompanied by a 'red shift' and an approaching light source by a 'violet shift.' Actually, my own feeling is that 'red shift' is a very poor term. It makes it sound as though the red light itself were shifting, or that light was shifting into the red. Neither is so.

Since any light of any wavelength would shift *toward* the red in case the source is receding, we ought to call it a 'redward shift.' The reverse would be a 'violetward shift,' or if that is too ugly a phrase 'blueward shift' would do almost as well.*

But now that we've decided the Doppler effect ought to apply to light as well as to sound, using pure reason, how can we play the game of science and test the theory by observation?

Why not repeat the trick that worked for sound? A locomotive whistle demonstrates a marked and unmistakable drop in pitch when that whistle passes us at a tenth the velocity of sound. Ought not a locomotive headlight (emitting a particular wavelength of light) demonstrate an equally marked and unmistakable change in color when it passes us at a tenth the velocity of light?

The trouble is that light travels so much faster than sound. Light travels at very nearly 300,000,000 meters per second. A tenth of that velocity is 30,000,000 meters per second and, let's face it, no locomotive can go that fast.

Suppose, then, we make do with what we have and deal with a locomotive traveling at the usual 33 meters per second, as in the case of sound. Here, though, the change in velocity is from $+ 33$ meters per second (approaching) to $- 33$ meters per second (receding) or 66 meters per second altogether. This is less than 1/4,000,000 the speed of light. The wavelength of light would shift by about this proportionate amount and detecting so tiny a shift would be difficult – prohibitively difficult in Doppler's time.

Doppler could scarcely be expected, then, to demonstrate a Doppler effect in light by purely Earth-bound experiments, despite his great success with respect to sound.

But what about the heavenly bodies? They moved much more rapidly than man-made objects did or could. The Earth, for instance, rotates in 24 hours, which means that a spot on Earth's equator is moving (relative to Earth's center) at a speed of about 465 meters per second, or 14 times the speed of an express train. The Moon revolves about the Earth at a speed of about 1000 meters per second. The Earth revolves

* I just mention this as a personal piece of pedantry. I'm sure the phrase will stay 'red shift.'

about the Sun at a speed of about 30,000 meters per second, while Mercury at its closest approach to the Sun reaches a speed of 57,000 meters per second.

It seems reasonable to suppose that the velocities of heavenly bodies, relative to one another, are generally in the thousands and ten thousands of meters per second and this offers a little more hope. Earth moves, relative to the Sun, at a velocity equal to 1/10,000 the speed of light. Not much, but a lot better than 1/4,000,000.

So let's not worry about light sources attached to locomotives, and consider, instead, the light emitted by heavenly bodies.

Of course, if we were dealing with a locomotive, we could possibly work up a headlight that would deliver a single wavelength of light and that would be easy to follow as it shifted this way and that. If we deal with heavenly bodies, we must take the wavelengths we get and we find that such bodies, whether they shine of their own light, as the stars do, or of reflected light as the planets do, deliver a broad band of wavelengths. That makes the matter of measuring a small shift much more complicated.

Doppler thought this added complication was not fatal. He assumed that a star delivered only visible light, spread across the spectrum from violet to red and that this spread was a more or less even one. If the star were receding from us, he imagined that all the wavelengths would grow longer and that the light would crowd into the red end of the spectrum, leaving the violet end empty. As a result, the overall light of the star would be distinctly redder in color. The greater the speed of recession, the more extreme the crowding into the red end, and the more distinctly red a star would become. On the other hand, if a star were approaching us, the light would crowd the other way and the star would become bluish in color.

Doppler was buoyed up in this belief by the fact that there were indeed stars which were redder than most (Antares, for instance, and Betelgeuse) and also stars which were bluer than most (Rigel, for instance, and Vega). Doppler suspected that Antares and Betelgeuse were receding from us at brisk velocities and Rigel and Vega were approaching us and that this it was that accounted for their colors.

116

Unfortunately, Doppler had started with a false assumption, and one which he ought to have known was false. (Well, let he among us who is without intellectual sin cast the first stone.) The fact is a star doesn't radiate *only* visible light from violet to red. It also radiates in the infrared, beyond the long-wavelength end of the visible spectrum, and in the ultraviolet, beyond the short-wavelength end. This was known since 1801.

When a star recedes from us, then, the light shifts redward without crowding up against an end-of-the-red barrier as Doppler had imagined. The very-long-wave red shifts into the infrared, and is replaced by the not-so-long-wave red, which is in turn replaced by the not-so-not-so-long-wave red, and on. At the other end, the very-short-wave violet which shifts toward the red does not leave an empty place at the short-wavelength end of the spectrum. It is replaced by the near ultraviolet which moves into the visible region. Precisely the reverse happens when a star approaches.

In other words, infrared and ultraviolet provide a kind of cushion and the visible portion of the spectrum does not change substantially with approach or recession of a star. The star's overall color does not change. What *does* affect the color is the star's temperature. A star hotter than our Sun is bluer in color (which is true of Rigel and Vega), while one that is cooler than our Sun is redder in color (as in the case of Antares and Betelgeuse).

(You see, then, that the author of the book I referred to in my introduction to this essay *still* had a Dopplerian notion of the redward shift – a notion now over one century out of date. Why read any of the book then?)

In 1848, the French physicist Armand Hippolyte Fizeau pointed out Doppler's error. He explained that it was no use trying to observe color changes. What one had to do was pick out some particular wavelength in the spectrum and somehow mark it. Then that particular wavelength could be observed as it shifted.

How could a particular wavelength be marked? Actually, there was a way, and Fizeau pointed it out.

In 1814, a German optician, Joseph von Fraunhofer,

discovered that the spectrum of sunlight was crossed by hundreds of dark lines, each one in a fixed position.

It might well be that each line represented a particular wavelength that was *absent* in sunlight as it reached the Earth, and if so, that absent wavelength might shift and be followed in the spectrum produced by stars other than the Sun. (The dark lines in the Sun are unshifted because the Earth does not move toward the Sun or away from it, but at right angles to its position, and a right-angle motion produces no shift. Thus the lines in the solar spectrum can be used as a reference against which to compare other spectra.)*

As a result of Fizeau's suggestion, the shift of wavelength with approach or recession of the light source is properly called the 'Doppler–Fizeau effect.'

Of course, Fizeau's suggestion remained nothing more than that till some way could be found to observe and measure the Doppler–Fizeau effect. It wasn't easy. In 1848, it was very difficult to get a visible spectrum out of starlight, so that its dark lines might be studied. Then when astronomers managed to accomplish this task, it quickly turned out that the spectrum of one star was often widely different in basic appearance from that of another. This meant it was uncertain whether spectral lines in one star could reasonably be compared with those of another and that shifts could therefore be converted into velocities.

But then, in 1859, the German physicist Gustav Robert Kirchhoff showed that light produced by heated elements was emitted only in certain fixed wavelengths. Light passing through the vapors of a particular element was selectively absorbed, with only certain wavelengths being subtracted from the light. Each element emitted and absorbed the same particular wavelengths and no others. No one element emitted or absorbed wavelengths identical to those emitted or absorbed by any other element. Each element, then, possessed a 'finger-print' in the spectrum.

* Because the Earth's orbit is not exactly circular, the Earth does approach the Sun slightly for half a year and then recedes from it the other half. The rate of approach or recession is something like 3 meters per second on the average – small enough to be ignored.

It seemed overwhelmingly probable that light produced at the solar surface, passing through the somewhat cooler atmosphere of the Sun, lost certain wavelengths through absorption by that atmosphere. The dark lines in the solar spectrum indicated, then, the chemical constitution of the solar atmosphere.

The overall spectrum might vary from one star to another in accordance with changes in temperature and chemical constitution, but individual spectral lines would be constant. A hydrogen line was a hydrogen line and an iron line was an iron line.

Astronomers turned to a study of the spectral lines with confidence then, and in 1868 the English astronomer William Huggins observed the spectrum of Sirius and detected a slight redward shift of a hydrogen line (as compared with the analogous line in the Sun's spectrum). He concluded that Sirius was receding from us at a speed of 40,000 meters per second. This figure was eventually corrected by more refined observations, but for a first try it was pretty good.

Such shifts, some redward and some blueward, were detected in other stars and astronomers were very pleased. And yet there was a weakness in the logic; the rules of the game of science demanded more.

Observe! Fizeau had decided that as a light source receded, the spectral lines *ought* to shift to the redward. Neither he nor anybody else had at that time actually observed such a shift take place in the case of a light source known to be receding.

Huggins, on the other hand, had observed a redward shift in the spectrum of Sirius and had decided that the star *ought* to be receding. This second 'ought,' however, depends entirely on the first 'ought,' and until the first 'ought' is demonstrated, the second 'ought' means nothing.

The game of science, if we are to play it rigorously, requires that we find some light source that we know is moving away from us, making use of firm evidence *other* than the redward shift. Then if that light source also shows a redward shift, we are in business.

With that in mind, let's consider the Sun. I have said earlier that it neither approaches us nor recedes from us, on the whole – but it does rotate on its axis. From a study of the motion of sunspots, it becomes quite clear that the Sun makes a complete rotation in 25 days and 1 hour (at its equator). The evidence is utterly convincing, depends on direct observation, and no one disputes it. The Sun's circumference is 4,400,000,000 meters, so any point on its equator must travel that distance in 25 days and 1 hour and must therefore move at a speed of 2000 meters per second.

This means that at one edge of the Sun's disc, the equatorial surface is approaching us at a speed of 2000 meters per second, while at the other edge it is moving away from us at 2000 meters per second.

From 1887 to 1889, the Swedish astronomer Nils Christofer Duner studied the spectrum of light coming exclusively from one edge of the solar disc and then exclusively from the other side. He did indeed find a measurable blueward shift in one case and a redward shift in the other. What's more, the shift was just the amount one would expect from a velocity (relative to ourselves) which had been determined by an utterly different and quite reliable method.

This established the Doppler–Fizeau shift and for an entire generation astronomers were entirely happy with it, especially since the picture of the heavens which it helped draw made a lot of sense and introduced no unsettling matters.

But, you know, they still weren't playing the game with the proper rigor. Even if we must agree that a velocity of recession causes a redward shift, we still have to ask ourselves if anything else can *also* cause a redward shift. If there is more than one cause for a redward shift, then how can we be sure which cause is operative in a particular redward shift?

The game of science never allows permanent immunity for those who break the rules and, in the 1910s, the matter of redward shifts turned up something so surprising as to reopen the whole question.

I'll get into that in the next chapter.

10
The Distance of Far

Alas, I have always been characterized by a certain naivety, and this characteristic was especially pronounced when I was young.

When I was nineteen, for instance, I was invited to visit a family that lived in a neighboring state. I was told the train station at which to get off. It never occurred to me to ask how to get to the house from the station. It never occurred to me to get a taxi at the station and let the taxi driver find the house. It never occurred to me to telephone the prospective host and have him come to get me.

The only thing I could think of doing was to ask the ticket taker at the station how to get to the street I wanted. He directed me up a certain road. Dubiously, I asked, 'How far do I have to walk?'

Brusquely, he answered, 'Far!'

I sighed, fixed my eye on the distant horizon, and began walking. I had walked several miles before I thought it worthwhile to ask someone for further directions, for now, I thought, I ought to be nearly there and I could get specific advice on reaching the precise house.

You're probably ahead of me. I had far overshot my mark and had to retrace most of my steps. When the ticket taker told me I had far to walk, I had neglected to ask even that most elementary of questions: 'How far is far?'

'How far is far?' was exactly the question that astronomers were asking themselves as the nineteenth century opened. They knew the stars were far, far away, but what was the exact distance of far?

The beginning of the answer came in the 1830s when it was found that even the nearest star was 4.3 light-years away (with 5.8 trillion miles to the light-year). Eventually, roughly a century later, it was discovered that our Galaxy of well over a hundred billion stars stretched out in a vast flat spiral about 100,000 light-years across.

Distance may lend enchantment as far as many things are concerned, but it is a pain in the neck to astronomers. The farther away a star, the fainter it is, the smaller its parallax, the less noticeable its proper motion.

This means that as a star grows more distant, it becomes ever more difficult to tell just how far the ever farther is. By the use of parallax, for instance, the first-exploited (and still the surest) method of measuring stellar distances, one can't probe outward beyond a hundred light-years or so – which confines us to the immediate neighborhood of our solar system.

Therefore, as the twentieth century opened, the prospects for exploring the Universe *beyond* our Galaxy, in terms of judging distances, seemed fairly hopeless.

To be sure, it wasn't at all certain that there existed anything beyond the Galaxy. The only possibilities in that direction were certain foggy patches in the sky called nebulae. Some of these nebulae were definitely inside our Galaxy, but others might possibly not be. These suspicious nebulae grew to be of particular, and increasing, interest in the opening years of the twentieth century.

The best hope for gaining knowledge of any kind concerning the very distant lay in those astronomic measurements that were independent of distance. The most important measurements of this sort were the shift in spectral lines caused by the radial velocity of some astronomic object; that is, its motion toward us (in which case the shift is toward the blue-violet end of the spectrum) or away from us (in which case it is toward the orange-red end).

The farther away a star, the dimmer it is, and the harder it is to obtain a spectrum out of its light, and the trickier it is to detect and measure the position of the spectral lines, and the more difficult it is to measure the shift of those lines. To that extent, radial velocity is harder to measure as distance increases. However, if the spectrum can be obtained at all, then the radial velocity can be measured with roughly equal accuracy regardless of distance. The farthest object capable of yielding a photographable spectrum with recognizable lines can have its motion toward us or away from us determined as precisely as the nearest.

In the latter part of the nineteenth century, radial velocities were measured for many stars. (Thousands of radial velocities are now known.) The figures for such stellar radial velocities fall into a relatively narrow range. For some stars, the radial velocity is virtually zero. (After all, some stars may be paralleling our own course or may be cutting directly across our own line of motion at right angles, so that, at the moment, they are neither approaching nor receding. At the other extreme, some stars have a radial velocity of as much as four hundred or five hundred kilometers per second relative to the Sun. Such extremes are exceptional. Most stars have radial velocities of between ten and forty kilometers per second, and there seems no preference between approach and recession. Some stars approach and some recede.

From radial velocities, certain conclusions can be drawn as to proper motion (that is, motion across the line of sight). Such proper motion can only be measured directly for the nearest stars and the radial velocity of a particular star has no necessary relationship to the proper motion of that star. Over a large number of stars, however, there is a statistical relationship and this can be used to gain a notion as to the true motion, in three dimensions, relative to our Sun.

When this is done it seems, at first glance, that the stars in the Galaxy are like a swarm of bees moving at random in all directions. Closer investigation made it seem that a certain amount of order was to be found. In 1904, a Dutch astronomer, Jacobus Cornelius Kapteyn, showed that the stars were moving in two streams, the motion of one being in the direction opposite to the other.

Then, in 1925, another Dutch astronomer, Jan Hendrik Oort, explained these streams in terms of the rotation of the Galaxy. In general, the farther an astronomical object from the gravitational center about which it rotates, the slower its orbital motion. In our solar system, the more distant a planet is from the Sun, the more slowly it moves in its orbit. In our Galaxy, the more distant a star is from the galactic center, the more slowly it rotates in its orbit about the center.

Stars more distant from the galactic center than our Sun is would move more slowly than the Sun. We would gain on them and they would seem to drift backward relative to ourselves. Stars closer to the galactic center than our Sun is would move more quickly and would drift forward. Hence, we would find two streams in opposite directions.

Radial velocities, then, turned out to be an extremely powerful tool, for they gave us a picture of the vast slow turning of the huge Galaxy on its axis; a picture we could scarcely have gained with such certainty in any other way.

Yet this was only the beginning.

The next stage in the dramatic victories of radial velocity began in 1912, when the American astronomer Vesto Melvin Slipher * measured the radial velocity of the Andromeda Nebula. This was one of those nebulae which some astronomers felt might be outside our Galaxy. It was, indeed, the only naked-eye nebula which labored under such a suspicion and might therefore very likely be the farthest object the human eye could see without aid.

Still, however far it might be, Slipher could get out of its light a spectrum marked by spectral lines. He could identify the spectral-line pattern and could measure how far that pattern had shifted from the normal position. He could tell the nebula's radial velocity with equal ease, whether it were outside the Galaxy or inside it.

The shift was blueward and Slipher concluded that the Andromeda Nebula was approaching the Sun at a speed of two hundred kilometers per second; a figure within the range of the radial velocities observed for astronomical objects generally. Interesting as an item in the astronomical data books but not particularly noteworthy!

Slipher's success led him to try to measure the velocity of other nebulae which resembled the Andromeda but were fainter and, therefore, presumably more distant. By 1917, he had succeeded in measuring the radial velocities of fifteen of them.

And, by then, he was rather puzzled. In the absence of any reason to expect anything else, scientists usually begin by

* Slipher died in 1969, at the patriarchal age of ninety-four.

supposing that any set of measurements will show a random distribution. In other words, if the radial velocities of a number of nebulae are measured, approximately half ought to be approaching us and half ought to be receding from us.

That proved not to be the case. Of the fifteen nebulae whose radial velocities were measured by Slipher, only two (the Andromeda and one other) were approaching. The other thirteen were all receding from us.

Worse yet, the recession was unexpectedly fast. The thirteen which were receding were doing so at an average velocity of 640 kilometers per second, an average well above the extreme radial velocity observed for any star.

If the nebulae were part of our Galaxy, these measurements were highly disturbing. Why should one set of objects within the Galaxy be receding from us almost in toto and be doing so with such high velocities, when nothing else within the Galaxy acted similarly?

This uniqueness in properties almost argued, in itself, for the extragalactic nature of the nebulae.

Fortunately, the question as to whether the nebulae were galactic or extragalactic did not have to depend on so subtle a point. In 1917, while Slipher was worrying about this, another American astronomer, Edwin Powell Hubble, was making use of a new one-hundred-inch telescope on Mount Wilson, in California. This telescope was powerful enough to resolve the till-then-featureless haze of the Andromeda Nebula and show that it was a collection of unprecedentedly faint stars – faint because of their great distance.

That was the final piece of evidence required to show that the Andromeda Nebula, and other similar objects were collections of stars far outside our Galaxy and were, indeed, galaxies in their own right. From that point in time, it became proper to speak of the Andromeda Galaxy, rather than of the Andromeda Nebula, and to distinguish our own home collection of stars as the Milky Way Galaxy.*

* If you are curious, the Andromeda Galaxy is currently thought to be about 2.2 million light-years from us and as far as those objects visible to the unaided eye are concerned, that is the ultimate distance of far.

This helped matters. It seemed reasonable that objects outside our Galaxy should act differently in some ways than objects inside our Galaxy. It was not utterly surprising that galaxies moved more rapidly relative to each other than did the stars within a given galaxy – just as you might not be surprised to learn that automobiles moved, on the average, more rapidly on highways outside cities than on the streets within cities.

That still left the disproportionately high number of galaxies that were receding from us – thirteen out of fifteen.

But then, perhaps, Slipher had just happened, by a fantastic run of luck, to pick those galaxies that were receding from us. Surely, if additional galaxies were studied, the recessions and approaches might even out.

The American astronomer Milton La Salle Humason took up the task. It wasn't easy. Slipher had naturally studied the brightest galaxies, those that would yield a spectrum with the least difficulty. Humason had to pass on to dimmer ones. He was forced to make photographic exposures of days at a time in order to pick up the spectra of the tiny dim patches of nebulosity that was all that could be made up of the more distant galaxies. The difficulties were considerable, but he managed.

To his astonishment, though, all he ever got were redward shifts! It seemed as though *all* the galaxies (except two of the closest) were receding. What was worse was that the redward shifts were enormous in size, representing velocities not of hundreds, but of thousands of kilometers per second. In 1928, Humason measured the redward shift of a galaxy called NGC 7619, and found this measurement to indicate a velocity of recession of 3800 kilometers per second.

What made everything worst of all was that, in general, the dimmer the galaxy (and, therefore, presumably the farther from us it was), the more rapidly it seemed to be receding.

This was a great deal for astronomers to swallow. To have the velocity of a galaxy depend upon its distance from *us* gives us far too much importance. Why should distance from little old us influence the manner in which a galaxy moves? Is there something about our Galaxy that repels other galaxies, and does this force of repulsion grow stronger with distance? For a while, Albert Einstein played with this notion, but no one has ever detected any force, attractive or repulsive, that grew stronger with distance, and the possibility was dropped.

It became necessary, then, for astronomers to take another and harder look at the redward shift. Remember it is the redward shift that is measured; that is hard observation that must be accepted. The conclusion from that, however, that a galaxy is receding is pure deduction and may be wrong. To be sure, since the mid-nineteenth century, astronomers had been taking it for granted that a redward shift meant a recession of the light source, but was that the only possible thing it could mean?

Light, after all, traveled over long, long distances in reaching us from the distant galaxies; distances much longer than those traversed from stars within our Galaxy. Perhaps something happened to light over the extralong distances that imposed a redward shift on it even though the light source (a galaxy, in this case) was stationary, or almost stationary, relative to us. This might mean that redward shifts could mean velocities of recession in connection with the stars inside our Galaxy, but something else altogether, just possibly, where the other galaxies were concerned.

Could it be, for instance, that very thin wisps of dust and gas between the galaxies, building up over millions of light-years, gradually absorbed some of the light traveling toward us? Perhaps it absorbed short-wave light preferentially, removing part of the blue-violet end of the spectrum preferentially, and leaving the galaxies redder than they would normally be.

Amateurs, considering the matter of the redward shift of the galaxies, sometimes come up with this idea (as did the author of the book I referred to at the beginning of the previous chapter). This is not surprising for, actually, the idea is valid as far as it goes. The light of distant galaxies *would* be reddened in this way, but only by light subtraction at the blue-violet end of the spectrum; *not* by any shift in wavelength. This effect, in other words, would produce a reddening, but *not* a redward shift of the spectral lines.

Well, then, suppose that light, as it travels through space over long distances, gradually loses its energy at a rate so slow it becomes noticeable only over intergalactic gaps. The wavelength depends upon the energy content of the light, so this means that as light travels over millions of light-years, its wavelength gradually increases. Every wavelength, including those occupied by spectral lines, shifts toward the red end of the spectrum. Naturally, the farther a galaxy, the more energy its light loses and the greater the redward shift. The beauty of this is that it accounts for the increasing redward shift with distance without granting our Galaxy some special importance it cannot have. The distance is important in itself.

This notion of 'tired light' (as it is usually referred to) has its difficulties, however. Unless you want to abandon the law of conservation of energy, which scientists are *extremely* reluctant to do, you must suppose that as the light gradually loses its energy, something else gains it. Astronomers, so far, cannot suggest the manner in which light energy is transferred over intergalactic distances in such a way as to produce a redward shift. The proper energy receiver is missing. (For instance, obstructing molecules in space will absorb a photon of light, but will not then necessarily reradiate a somewhat less energetic photon *in the same direction* as the original photon was traveling. Gas and dust will absorb or scatter light but will do nothing more, and something more is required to fit the observations.)

Furthermore, the loss of energy by light, if sufficient to account for the redward shift in galaxies, is also sufficient to be detectable in intragalactic observations, and it isn't.

The tired-light hypothesis is, therefore, found wanting both in theory and observation and it must be (usually reluctantly) abandoned – at least until further information in its favor may happen to be adduced.

But here's something else. In 1916, Einstein advanced his General Theory of Relativity and pointed out that light moving against the pull of a gravitational field loses energy (without violating the law of conservation of energy). Light radiating outward from any star is moving against the pull of the gravitational field of that star so the light from any star or galaxy should show a gravitational redward shift.

Is it possible, then, that the redward shifts of the galaxies are gravitational in origin, rather than recessional?

The trouble is that under ordinary circumstances this shift is so small as to be all but undetectable. In order to make the gravitational redward shift large enough to be detected, a large gravitational field is in itself insufficient; it must be intense. A sufficiently intense gravitational field is produced only by large quantities of matter condensed to superdensity – by a white-dwarf star, for instance.

To suppose, then, that the redward shifts of the distant galaxies are gravitational in origin is to suppose fantastic densities for them. Even if this were swallowed, those densities would have to increase steadily with distance from us, and it is even more difficult to see why our location should affect the density of a distant galaxy than its velocity.

This brings us back to velocity of recession as the only reasonable explanation of the redward shifts, and to the puzzling relationship between this velocity and distance from us.

Hubble tackled the matter. He made use of every possible method for determining the relative distances of the galaxies. It is possible to detect in a few of the nearer ones certain pulsating stars called Cepheids. From their rate of pulsation and their apparent brightness, their relative distances (and, therefore, the relative distances of the galaxies containing them) could be determined.

In more distant galaxies, Cepheids could not be made out but a few extremely luminous stars could be seen. Assuming that there is some limit to luminosity and that the most luminous stars in each galaxy are at this limit and are therefore roughly equal in luminosity, the relative distances of the galaxies containing them can be determined.

Finally, where galaxies are too distant to reveal any stars at all, it may be assumed that their total luminosities are roughly equal and, from their apparent overall brightness, their relative distances may be determined.

Once this was done, the velocity of recession, as measured by the red shift, comes so close to bearing a direct relationship to the distance of the galaxy from ourselves that, in 1929, Hubble announced this direct relationship as really existing. This is called 'Hubble's Law.' If galaxy A is x times as far from us as galaxy B, then galaxy A is receding from us x times as quickly as galaxy B.

Unexpectedly, then, if Hubble's Law is correct, astronomers suddenly had a very powerful tool for measuring the distances of even the farthest visible objects. Once the distances of the nearer galaxies could be estimated by some method (any method) not involving the redward shift, then the distance of any farther galaxy was instantly known.

In the 1950s the 200-inch telescope could detect galaxies which were probably up to 1.5 billion light-years distant, and in the 1960s quasars were discovered with distances from us of up to 8 or 9 billion light-years, while the edge of the observable Universe can be calculated as being 12.5 billion light-years distant.

But still the reason for the relationship between distance and recessional velocity eludes us.

The answer came from Einstein's General Theory. In it, Einstein had worked out a set of 'field equations' which described the overall properties of the Universe. (This was the beginning of modern cosmology.) Einstein solved the field equations in such a way as to picture a static Universe, one in which the overall density of matter remained constant.

In 1917, however, the Dutch astronomer Willem de Sitter pointed out that another solution was possible, one in which the overall density of matter in the Universe was constantly decreasing with time.

One way of imagining such a constant decrease of overall density is to suppose that the Universe consists of particles of matter of fixed density which are forever moving apart from each other at constant velocity. The Universe would then consist of the unchanging particles plus more and more interparticular space and the overall density would go down.

De Sitter worked out his solution as a purely theoretical exercise, but when Hubble worked out his law, it did not take long to see that that law was a consequence of the De Sitter suggestion.

In the Universe, the individual galaxies may be considered as particles. They are held together by the mutual gravitational attraction of their constituent stars, so that the overall density within a galaxy remains unchanged with time. In fact, a number of galaxies, relatively close to each other, can remain bound together by gravitational forces, so that the overall density within galactic clusters will remain unchanged. When I speak of galaxies in the following paragraphs, then, let it be understood that I refer either to isolated galaxies or to gravitationally bound clusters of galaxies.

If the galaxies move steadily apart, the overall density of the matter in the Universe constantly decreases. We are then picturing an 'expanding Universe.'

In a constantly expanding Universe, an observer on some one of the galaxies will see all the other galaxies receding. Furthermore, it is easy to show (though I won't try to do so here) that in such a Universe Hubble's Law must hold. The farther a galaxy from the observer's Galaxy, the faster the velocity of recession of the galaxy relative to the observer's Galaxy.

This removes the apparent paradox from Hubble's Law. There is no magic about us, no queer influence in our Galaxy that relates recessional velocity to distance from *us*. What we see, we would also see from any other galaxy in the Universe.*

It is startling to think, then, that what began with an Austrian physicist listening to trumpet players sounding notes as they moved past him on a railroad flatcar (see Chapter 9) ended less than a century later by producing a grand vision of a Universe billions of light-years across engaged in a steady and colossal expansion.

It is this kind of procession from the utterly prosaic to the unimaginably ultra that is what can happen when the game of science is played correctly.

* Nor is the approach of the Andromeda Galaxy a violation of the principle of the expanding Universe. The Andromeda Galaxy is part of a cluster of galaxies which also includes our own Milky Way Galaxy. The two galaxies along with some two dozen 'dwarf' galaxies are gravitationally bound, and move about relative to each other independently of the general expansion of the Universe.

C
Chemistry

11

The Multiplying Elements

When, as a schoolboy, I found either the teacher or the subject (or both) dull, my attention would naturally wander. There is nothing as painful as boredom and a wandering attention is thankful for even the slightest relief.

Chemistry was, in that respect, wonderful, for I have never been in a decent chemistry lecture room in which a large periodic chart of the elements was not hanging somewhere in the front of the room. One could study it without too obviously turning one's eyes away from the lecturer and it was complicated enough to allow hours of contemplation.

The old-style charts of the late 1930s charmed me most with the manner in which a whole group of elements tried to squeeze their way into a single slot. The chart had to have an asterisk at that point, and the elements had to be spread out at the bottom. They were labeled the 'rare-earth elements.'

While I was still relatively naive about the periodic table, this little item afforded me much speculation and preserved my sanity when the tide of boredom washed dangerously near my mental nostrils. Why 'rare'? I would wonder. Why 'earth'? I would wonder. Why do they all insist on crowding into the same spot? I would wonder.

I eventually found the answers to the questions and you know me: I like nothing better than sharing these little finds with you.

The ancient Greeks* considered one of the basic components of the Universe (i.e., 'elements,' to use the Latin term) to be 'earth.' By this, they didn't really mean the literal earthy stuff we stand on, but a kind of ideal solid which was to be found in differing proportion in the various components of the planetary crust.

* I love to start an essay with 'The ancient Greeks' – and I frequently do.

134

Some of those components were rather unearthy. There were, for instance, the various metals, which had luster and were malleable, whereas typical earthy components were dull and brittle. There were also carbon and sulfur, which lacked luster and were brittle, but which burned, whereas typical earthy components were unaffected by heat.

We might say that substances which are sufficiently earthy to be considered earth ought to be dull, brittle and unaffected by heat – like the rocks, sand and clay that are all about us.

Even if we confine earth to substances with those properties, it still turns out that the term applies not to a single material but to a large group of minerals that differ among themselves. There was no earth, but merely different 'earths.'

Using modern terminology, we would define an earth as a stable oxide with a high melting point. The four most common of these were silica, alumina, lime and magnesia, in that order. In modern terminology, we would call them silicon dioxide (SiO_2), aluminum oxide (Al_2O_3), calcium oxide (CaO) and magnesium oxide (MgO). Singly and in combination (as 'silicates'), these four earths make up just about two thirds of the earth's crust. It was impossible to pick up any sizable chunk of that crust anywhere without finding quantities of each of the four.

Chemists of the late eighteenth century suspected that these earths were metallic oxides, and in due time those metals were isolated. (Silicon is at best a semi-metal, I hasten to say, before the letters pour in.)

Two other very common solid substances, obtained in one fashion or another from the environment, were soda and potash. In modern terminology, these are sodium carbonate (Na_2CO_3) and potassium carbonate (K_2CO_3). Chemically, they differ from the earths in that they are carbonates rather than oxides but that was not the distinction that forced itself upon the early chemists.

What was more important to them was that soda and potash were freely soluble in water, something that was not true of an earth. Then, too, soda and potash, once dissolved, have distinctive properties, such as the ability to neutralize acids.

Since soda and potash were most easily obtained by burning

135

certain plants in big pots and extracting the ash (hence 'pot-ash'), they were called alkalis, from an Arabic word meaning 'ash.' (The Arabs were the great alchemists of the early Middle Ages, and chemistry went through a period in which great prestige was attached to Arabic words.) Solutions of soda and potash were said to be alkaline in properties.

Well, it so happens that of the very common earths, two – lime and magnesia – are somewhat soluble in water and produce solutions with alkaline properties. They were therefore referred to as the 'alkaline earths' and are still so named to this day. The metals obtained from lime and magnesia (calcium and magnesium, respectively), plus similar metals isolated from related, but less common, earths (beryllium, strontium and barium) were, and are, termed the alkaline earth elements.

Indeed, when, in the last years of the nineteenth century, the glamorous element, radium, was discovered, it turned out to be a member of the same family – very similar to barium in its chemical properties, in fact. Radium is an alkaline earth metal.

(Have no fears that the soil will wash away because two of the important earths are soluble. Lime and magnesia occur in combination with silica, as calcium and magnesium silicates, and in that form are quite insoluble.)

Now we are ready to go on to the next part of the story.

Three miles from Stockholm was a quarry in a hamlet which bore the name of Ytterby ('outer village').

One day in 1787, a Swedish army officer, Lieutenant Carl Axel Arrhenius (1757–1824), who was an amateur mineralogist, picked up an unusual black rock at the quarry. He could not identify it and rightly assumed it was a mineral that had not been scientifically studied hitherto. He called it 'ytterbite,' making use of the 'ite' suffix commonly given to minerals.

The curious mineral attracted attention, of course, and in 1794, a Finnish chemist (in those days, Finland was part of the Swedish realm) named Johan Gadolin (1760–1852) took it apart chemically. It was a combination of several different oxides, one of which was silica.

One of the oxides, which Gadolin separated out and found to

make up nearly two fifths of the whole, struck him as most unusual. It was not to be identified with any other known substance and yet it had all the properties of an earth. It was insoluble, non-metallic and unaffected by heat. He announced it as a new earth and called it 'yttria.' It was clear by that time that earths were oxides and a new earth would have to be the oxide of a new metal. The actual metal of this earth was not isolated for another half century, but that was a detail. It was there; it was known to be there; and the discoverer of the earth was considered the discoverer of the metal. The metal was almost invariably named after the earth in the case of new discoveries, with the 'ium' ending conventionally added. Thus yttria was the oxide of 'yttrium' and Gadolin was considered its discoverer.

In 1812, a Scottish chemist, Thomas Thomson (1773–1852), visited the quarry and marveled over the mineral. He was still under the spell of the word 'earth.' If an object was called an earth, surely it would have to be a major component of the earth. The common earths were all major components of the earth's crust and were found everywhere, so surely that proved that that was the way an earth *ought* to be. (I'm not responsible for this curiously circular logic; this sort of thing often happens when human terminology is mistaken for natural law.)

Now Thomson was staring at an earth, yttria, that could only be found in one or two favored spots and was unknown elsewhere. He wrote, concerning it, 'A peculiar earth confined to a peculiar spot and in very minute quantities, can hardly be conceived.'*

Yttria, in other words, was a 'rare earth' which, to Thomson, was almost a contradiction in terms, and that very contradiction helped to make the phrase notable to chemists, generally. Thomson cited two other examples of rare earths: glucina (beryllium oxide, BeO) and zirconia (zirconium oxide, ZrO_2). However, because of developing events, the title of 'rare earths' was not applied to all earths which happened to be rare, but was

* I refer you, by the way, to a book called *Discovery of the Elements* by Weeks and Leicester, seventh edition (Journal of Chemical Education, 1968). It is loaded with curious information scarcely to be found conveniently elsewhere.

restricted to yttria and its relatives – for indeed yttria turned out to have relatives.

Gadolin lived a long and respectable life, serving with honor as a professor of chemistry at the University of Abo, but one must admit that his discovery of yttria was his only deed of real note – and yet it was enough. What Lieutenant Arrhenius had called 'ytterbite' was renamed 'gadolinite' in his honor, and more awaited him in the future.

Once Gadolin had made his splash, other chemists naturally yearned to get into the act and Swedish minerals began to get a fine-tooth combing. One of the chemists who plunged into action was a German named Martin Heinrich Klaproth (1743–1817). He had already made a name for himself as a discoverer of elements. In 1789, he had discovered zirconium (whose earth was listed by Thomson as one of the apparently paradoxical rare earths, but which isn't included in that term today) and uranium. In 1798, he discovered titanium.

Now, in 1803, he tackled a heavy Swedish mineral that was not quite like gadolinite but that looked promisingly novel. From it, he isolated an earth that was not quite like yttria and yet was not like anything else, either. Obviously it was something new and he called it 'terre ochroite,' meaning 'pale yellow earth,' which, I think you will agree, is a perfectly rotten name.

At almost the same time, Sweden's (and Europe's) greatest living chemist, Jöns Jakob Berzelius (1779–1848), working with a Swedish geologist, Wilhelm Hisinger (1766–1852), isolated the same earth. Klaproth was first by a hair, so he gets the credit for the discovery, but it was Berzelius who gave the new earth a sensible name, and, oddly enough, he used a precedent established by Klaproth.

In the Middle Ages, the alchemists had named the various metals after the various planets. When Klaproth had discovered a new element in 1789, he remembered that only eight years before a new planet had been discovered for the first time in the period of written history. The new planet had been named Uranus and it seemed appropriate to Klaproth to name the new element 'uranium' in its honor.

When Berzelius and Hisinger obtained their new earth in

1803, they kept in mind the fact that two years earlier anot...
new planet was discovered and that had been named Ceres.
They called their earth 'ceria' therefore, and it was that name
which stuck. (Ceres was the precursor of a whole family of
similar planets – the asteroids – and cerium, with yttrium,
was the precursor of a whole family of similar elements. This is
the kind of coincidence that wouldn't be believed in a work of
fiction.)

For something like a generation, the two sister earths (for
they were very similar in chemical properties), yttria and ceria,
were accepted for what they were and then one chemist, at
least, began to wonder. The two earths were sufficiently similar
so that, if mixed, they would be hard to separate by nineteenth-
century techniques. Could one be sure that *other* earths were
not mixed with one or the other or both?

The chemist who wondered about this was Carl Gustav
Mosander (1797–1858), a pupil of Berzelius. He had already
involved himself with the rare earths when he tried to isolate
the metallic portion of ceria by using potassium vapor to snatch
away the oxygen. He managed an imperfect separation by this
means (and handling potassium vapor isn't something I would
ever care to try myself) and obtained an impure sample of the
metal, cerium. He was the first to isolate a rare earth metal, but
really pure samples of cerium and its sister metals had to wait
for the twentieth century.

Now in search of new earths, he treated ceria with nitric acid
and found, sure enough, that some of it dissolved more easily
than the rest. The dissolved portion he separated and found it
to be an earth that was not ceria, though very like it. He called
this new one 'lanthana' from a Greek word meaning 'hidden'
because it was, after all, hidden in the ceria. (The name was
Berzelius' suggestion.)

But Mosander wasn't finished. He kept working with the
lanthana he had obtained to see if *it* was pure. It took him two
years of painstaking treatment by the crude methods of those
days, but by 1841, he had satisfied himself that the lanthana he
had obtained from ceria was itself not pure but contained
another earth that was still more subtly hidden. This new
earth he called 'didymia' from the Greek word for 'twin'

139

because it seemed to him to be an almost inseparable twin of lanthana.

This had gone past a joke by now. One unusual earth, yttria, had been remarkable and enough. To have it multiply like rabbits was disturbing. Four similar earths now existed: yttria, ceria, lanthana and didymia, in that order of discovery. Who was to say how many more there might be?

The term 'rare earths' began to be applied to this family specifically.

Mosander continued his disturbing way, too. He had tackled Klaproth's ceria and had ripped two new earths out of it. Wasn't it appropriate that he do the same for Gadolin's yttria, which was now half a century old and had not yet had a proper ransacking?

He began his treatment of samples of yttria. Carefully, he removed from it any ceria, lanthana and didymia it contained (and it did contain them, to be sure, for any sample of a rare earth, prepared without unusual care, seemed to contain bits and pieces of all the family).

Once that was done, he ought to have had pure yttria, but did he? By careful treatment with nitric acid, he finally obtained, in 1843, three different earths of different solubility in the acid. Each even had its own color. A colorless earth made up the largest fraction of the original and for this he retained the name 'yttria.' A yellow earth, which he isolated from the original, he called 'erbia' and a rose-colored one, 'terbia.'

All three names, yttria, erbia and terbia, are derived from the name of the original quarry, Ytterby. It makes sense to do this in one way, since it shows the close chemical relationship of the three earths and their common origin.

In another sense, however, it pays entirely too much honor to Ytterby. The number of elements is limited and their names should be carefully conserved for appropriate commemoration. (Of course, they weren't; many of the names were chosen for utterly trivial reasons and it is too late now ever to do anything about it.)

What's more, the names erbia and terbia are entirely too similar in sound and it is absurdly easy to confuse them. Indeed, Mosander's original assignment of the two names was

140

twisted and what he originally called erbia came to be called terbia and vice versa. This made no real difference but it shows the complication that can ensue when names are carelessly chosen.

There the situation stood for another generation. There were now six rare earths and they were a standing embarrassment to chemists. In 1869, Dmitri Ivanovich Mendeléev (1834–1907) devised a periodic table of the elements, which arranged the elements in a rational order based on their chemical properties. The usefulness of this table was shown by the fact that it made it possible to predict certain elements that remained to be discovered and to describe their properties with a high degree of accuracy. All the elements ought to have fit neatly into the table and two of the rare earth elements known at the time – yttrium and lanthanum – did. The other four, however, didn't and that part of the situation grew rapidly worse. Other rare earths were discovered and all but one did not fit into the table.

In 1878, a Swiss chemist named Jean Charles Galissard de Marignac (1817–94) tackled erbia (the rose-colored earth which Mosander had originally called terbia). By further treatment he divided it into a colorless earth, and one that was a darker red than the original. Since the original erbia was colored, he kept the name for the red earth. The other one, colorless, he called 'ytterbia,' *also* from that notorious Ytterby quarry, the *fourth* earth (and, therefore, element) to be so honored.

The very next year, 1879, a Swedish chemist, Lars Fredrik Nilson (1840–99) began with Marignac's ytterbium and managed to get still another earth out of it. Fortunately, he couldn't think of a fifth way to torture Ytterby into supplying a name, so he generalized and called it 'scandia' for the Scandinavian peninsula as a whole.

This earth *did* fit into Mendeléev's table (the last of the rare earths to do so where the nineteenth-century version of the table was concerned) and in a particularly glamorous way. When Mendeléev had worked out his table, he used it to predict the properties of three elements that were as yet undiscovered. One of those elements had been discovered in 1875 and named 'gallium' and had fit the prediction perfectly. The new earth, scandia, contained the metallic element 'scandium,' which was obviously the equivalent of the second of Mendeléev's predicted elements and it fit the predicted properties perfectly also.

But if Marignac's ytterbia had yielded a new earth, what about the erbia from which that ytterbia had come? Could the erbia be made to yield still more goodies?

Another Swedish chemist, Per Teodor Cleve (1840–1905) found it could. In 1879, he completed the separation of erbia into three fractions. For the major fraction he retained the name erbia. The two minor fractions he called 'holmia' and 'thulia.' Holmia was named for Stockholm, the city in which Cleve was born. Thulia was named for Thule which, in ancient times, was the legendary land of the far north and which some people equated with Scandinavia.

And still it went on, when a French chemist, Paul Emile Lecoq de Boisbaudran (1838–1912), took a hand. It had been he who had discovered gallium by means of newfangled spectroscopic techniques. It turned out that each element gave its own combination of spectral lines when heated white hot, and Lecoq now turned his spectroscope on the rare earths.

It would no longer be necessary to work blindly. If two samples of a particular earth, obtained from different minerals, or prepared by different techniques, showed differences in their spectra, then the sample wasn't pure and more than one substance was present.

Lecoq found Mosander's old earth, didymia, to be spectroscopically suspicious and began to fractionate it, being guided by spectra at every step. In 1879, he split off a new earth and named it 'samaria.'

One might think that this name comes from the biblical city of Samaria, but it does not. It happened that the ore Lecoq was working with was named 'samarskite.' What's more, samarskite was a type of ore first discovered in Russia and was named after one Colonel Samarski, an obscure mining official. Samarium was the first element to be named for a human being and, as luck would have it, it was for someone of no recognizable merit whatever.

Lecoq did better in 1886 when he obtained still another earth from samaria, identical to one that Marignac had earlier obtained (with less certainty) in 1880. Marignac gets the credit for the discovery but it was Lecoq who gave it its name. With Marignac's permission, he named the new earth 'gadolinia' after old Johan Gadolin, who had started the whole thing nearly a century before and who was now dead a generation. That at least was worthy.

Samarium and gadolinium are the only two stable elements in the entire list of eighty-one to be named for human beings. Other and greater men have been honored but names such as Mendeléev, Curie, Einstein and Fermi have had to be used for short-lived radioactive elements discovered after World War II. (Those radioactive elements have, by the way, an interesting connection with the rare earths, but that is another story for another time.)

In the space of a little over two years, then, no less than six new rare earths had been found and there was *still* no way in which chemists could predict how many more remained to be discovered.

Didymia, for instance, which had served as source for samaria and gadolinia, still looked suspicious even after those two had been separated out. An Austrian chemist, Karl Auer, Baron von Welsbach (1858–1929), tackled didymia in 1885 and found that the earth, which had originally been named 'twin,' was indeed a pair of twin elements. He split it neatly in two.

For the first time, a separation did not leave one preponderant fraction to which the original name could still be applied. Didymia disappeared altogether from the list of earths, the only established rare earth (it had been on the list for forty-four years) to do so. The two fractions were named 'praseodymia' ('green twin' because it was a greenish yellow in color) and 'neodymia' ('new twin').

The next year, 1886, Lecoq de Boisbaudran, worked with Cleve's 'holmia' and managed to isolate still another rare earth, which he named 'dysprosia' after a Greek word meaning 'hard to get at.'

Things slackened off after that but did not die altogether. Even after the opening of the twentieth century, new earths were found. In 1901, a French chemist, Eugène Demarçay, like

Lecoq a crackerjack spectroscopist, put samaria through a series of elaborate fractionations and ended with a new earth he named 'europia' after the entire continent.

Finally, in 1907, another French chemist, Georges Urbain (1872–1938), did the same to ytterbia and isolated still another rare earth which he used for a far more specific honor. He called it 'lutetia' after the Roman name of the town which later came to be called Paris. (Urbain was born in Paris, needless to say.)

Here let us call a temporary halt and take a look at the rare earths that were known up to 1907. I will list them in the order of discovery:

Rare earth	Metal	Year discovered	Discoverer
yttria	yttrium	1794	Gadolin
ceria	cerium	1803	Klaproth
lanthana	lanthanum	1839	Mosander
didymia	didymium	1841	Mosander
erbia	erbium	1843	Mosander
terbia	terbium	1843	Mosander
ytterbia	ytterbium	1878	Marignac
scandia	scandium	1879	Nilson
holmia	holmium	1879	Cleve
thulia	thulium	1879	Cleve
samaria	samarium	1879	Lecoq de Boisbaudran
gadolinia	gadolinium	1880	Marignac
praseodymia	praseodymium	1885	Welsbach
neodymia	neodymium	1885	Welsbach
dysprosia	dysprosium	1886	Lecoq de Boisbaudran
europia	europium	1901	Demarçay
lutetia	lutetium	1907	Urbain

Seventeen rare earths were discovered. To be sure, didymium disappeared from the list but that still leaves sixteen.

Sixteen elements that formed a tight family, very similar in chemical properties, impossible to separate completely by nineteenth-century techniques! Chemists still hadn't the faintest idea how many more might be found and they were a major embarrassment.

Mendeléev's periodic tables was inadequate to help in this respect. Three of the rare earth elements – scandium, yttrium and lanthanum – could be placed: the rest had nowhere to go.

But if the rare earths were a pain in the neck, it is just such pains in the neck that grease the wheels of scientific progress. If the periodic table was to be satisfactory, it would *have* to account for the rare earth elements, and that meant something more fundamental than anything the nineteenth century had thought of would have to be added to it.

Fortunately, the necessary something was found, but first let's consider the periodic table in the next chapter.

12

Bridging the Gaps

In 1969, Houghton Mifflin published my hundredth book.*
The Boston *Globe* went on to celebrate the occasion with a long
article and the New York *Times* followed with another article.
What's more, Houghton Mifflin threw me a cocktail party on
publication day.

All in all, this was enough to turn anyone's head so, lest I
lose the lovable modesty which is my hallmark, I am keeping
firmly in mind something that once happened to my mother.

Back in the early 1950s, my parents finally sold their candy
store and moved into well-earned retirement. Naturally, time
hung heavy on their hands, so my father got a part-time job
that only took up forty hours a week (the candy store had taken
up ninety) and my mother went to night school.

My mother felt keenly her inability to write English. She
could write Russian and Yiddish, but neither language used the
Latin alphabet. She could read English but didn't know the
written script; so she took a course in writing and made
marvelous progress. In no time at all she was writing me letters
in clear script.

Then, one evening she was stopped in the hall by a member of
the night-school faculty who proceeded to ask her what we, in
our family, call That Good Old Question. He said, 'Pardon me,
Mrs. Asimov, but are you related to Isaac Asimov, by any
chance?'

My mother at once said, 'Yes, indeed. Isaac Asimov is my son.'

The teacher said, 'Oh! Then no wonder you're such a good
writer!'

Upon which my mother, well-aware of the unidirectional
flow of genes, drew herself up to her full four-feet-ten and said,

* The name is *Opus 100* to forestall questions, and it is by way of
being a kind of literary autobiography, with illustrative selections from
earlier works.

freezingly, 'I *beg* your pardon, sir. No wonder *he's* a good writer.'

And, relying on the memory of that chastening remark to produce the proper sense of humility within myself, I will now turn to my subject – which will continue from where I left off in the last chapter.

In the mid-nineteenth century, some three-score elements had been discovered and chemists were growing rather anxious. Each decade was seeing the number increase: three had been discovered in the 1770s, five in the 1780s, five in the 1790s, *fourteen* in the 1800s, four in the 1810s, five in the 1820s, and so on.

Where would it all end? Scientists value simplicity and when what had seemed simple to begin with grows increasingly complex, a new order of simplification is sought for. In this case it grew tempting to find some way of ordering the tangled list of elements so that 'element families' might show up. This would tame the jungle somewhat.

Indeed, if the elements were properly arranged, there might even be some way of determining how many elements might exist altogether and, therefore, how many remained yet to be discovered. In the mid-nineteenth century, however, this seemed like a mighty far-out thought.

The one quantitative measurement known for the atoms of the various elements at that time was the atomic weight. Thus, if the weight of the hydrogen atom (the lightest known, both then and now) is considered to be 1, the carbon atom, twelve times as massive, is assigned an atomic weight of 12, the oxygen atom is 16 and so on.

To begin with, then, one might try to arrange the elements in the order of atomic weight to see if some rational family system would then show itself. It turned out that a rectangular table *could* be designed in which similar elements would occasionally show up in rows or in columns (depending on whether successive elements are arranged vertically or horizontally). Unfortunately, the earliest tables also put together some very *un*like elements, and in science a half solution is no solution at all.

The basic trouble, you see, with atomic weights as a guide for element arrangement is that there is no way of knowing when the list is complete. It happens that the atomic weight of carbon is 12, of nitrogen, 14 and of oxygen, 16. How can you be sure that there are not some as yet undiscovered elements with atomic weights of 13 and 15? The entire table might be thrown out of whack simply because missing elements are not included.

You might argue, of course, that a difference of 2 in atomic weight was as close as elements are likely to get, but you can't be certain of that. Nickel has an atomic weight of 58.7 and cobalt one of 58.9. With such atomic weight differences there would be room for nine elements between carbon and nitrogen and nine more between nitrogen and oxygen.

The fact of the matter is simply that atomic weights are not enough. Some other quantitative properties are required and, best of all, some properties which can be presented only as integers so that if you get a 1 and a 2 and a 3, you know there can be nothing in between.

The beginning of the breakthrough came in 1852. An English chemist, Edward Frankland, took note of the fact that in the chemical formulas that were being worked out an atom of a particular element seemed always to combine with a fixed number of other elements.

Thus, a hydrogen atom never combined with more than one other atom. It could be assigned a combining power (or 'valence' from a Latin word meaning 'power') of 1. An oxygen atom could combine with two hydrogen atoms, a nitrogen atom with three and a carbon atom with four, so that oxygen, nitrogen and carbon had valences of 2, 3 and 4 respectively. These valences worked out very nicely. Thus a carbon atom (valence, 4) could combine with two oxygen atoms $(2 + 2)$ or with one oxygen atom and two hydrogen atoms $(2 + 1 + 1)$.

The valence concept not only had the virtue of simplicity and of clear and evident usefulness, it also offered the required integral property, for there seemed no possibility of valences of 1.5 or 2.32 or anything like that.*

* Actually, twentieth-century sophistication introduced new concepts of valence that did indeed involve something like fractional values but that does not affect the line of argument in the chapter.

In 1869, the Russian chemist Dmitri Ivanovich Mendeléev tried arranging the elements according to molecular weight *and* valence. The result was a table of which I will present a very simplified and incomplete version (see Table 1) with the atomic weights given to one decimal place.

In Table 1, I am using the chemical symbols for the elements in order to save space, but that doesn't affect the argument or in any way confuse it even if you don't know what elements the symbols stand for. When I have to mention a particular element, however, I will give the full name as well as the symbol.

The rows in Table 1 do indeed contain closely knit element families. For instance, the top row contains lithium (Li), sodium (Na), potassium (K), rubidium (Rb), cesium (Cs) and francium (Fr), which all have very similar properties. All are low-melting, exceedingly active metals which, under given chemical conditions, react in nearly identical fashion. What's more, where differences do exist, they show a steady gradation as one goes across the row. Moving from lithium to sodium to potassium and so on, we find that the melting point gets progressively lower and the activity progressively higher. These six are the 'alkali metals.'

The second row contains six 'alkaline earth metals' which are also similar among themselves, and so on.

Notice that in Period 5, tellurium (Te) comes before iodine (I), even though tellurium has the larger atomic weight and therefore ought to come after iodine if atomic weight were the sole criterion.

It was one of Mendeléev's great decisions to place questions of valence (and chemical properties in general) ahead of that of atomic weight. In order to place tellurium and iodine in the proper family with the proper valence, the atomic weight order had to be inverted. The more sophisticated knowledge of atomic structure gained by later chemists proved that in this respect Mendeléev's intuition was absolutely correct.

As one goes down the list of elements according to molecular weight, a particular set of properties turns up periodically; therefore the listing, when arranged so that those particular sets fall into neat rows or columns, is called a 'periodic table.'

At the time Mendeléev first advanced the periodic table, a number of the elements given in Table 1 were not yet discovered. These are indicated in Table 1 by asterisks.

As an example, the six elements in the bottom row, helium (He), neon (Ne), argon (Ar), krypton (Kr), xenon (Xe), and radon (Rn) were all unknown in 1869. Their existence was

TABLE 1 – THE VALENCE ELEMENTS

Valence	Period 1	Period 2	Period 3	Period 4	Period 5	Period 6	Period 7
1		Li	Na	K	Rb	Cs	Fr*
		(6.9)	(23.0)	(39.1)	(85.5)	(132.9)	(223)
2		Be	Mg	Ca	Sr	Ba	Ra*
		(9.0)	(24.3)	(40.0)	(87.6)	(137.3)	(226)
				#	#	#	
3		B	Al	Ga*	In	Tl	
		(10.8)	(27.0)	(69.7)	(114.8)	(204.4)	
4		C	Si	Ge*	Sn	Pb	
		(12.0)	(28.1)	(72.6)	(118.7)	(207.2)	
3		N	P	As	Sb	Bi	
		(14.0)	(31.0)	(74.9)	(121.8)	(209.0)	
2		O	S	Se	Te	Po*	
		(16.0)	(32.0)	(79.0)	(127.6)	(210)	
1	H	F*	Cl	Br	I	At*	
	(1.0)	(19.0)	(35.5)	(79.9)	(126.9)	(210)	
0	He*	Ne*	Ar*	Kr*	Xe*	Rn*	
	(4.0)	(20.2)	(39.9)	(83.8)	(131.3)	(222)	

utterly unsuspected and the periodic table seemed complete without them. As one goes up the list of atomic weights, the valence change in the elements listed in Table 1 (minus the bottom row) would be 1,1,2,3,4,3,2,1,1,2,3,4,3,2,1,1,2 and so on.

When the bottom-row elements were discovered, however, it was found that they did not combine with any other elements and therefore had a valence of 0.* The valence change was therefore 1,0,1,2,3,4,3,2,1,0,1,2,3,4,3,2,1,0,1,2 and so on.

The bottom-row elements, all very similar in properties and called the 'inert gases' or the 'noble gases,' therefore extend the table without upsetting it. Quite the contrary, the insertion of a

* Exceptions to this general rule were discovered in the 1960s but these are exceptions. The valence of 0 under ordinary conditions stands.

0 in the proper place makes the periodic table even more elegant. The fact that an unsuspected group of elements should so remarkably fit and beautify a periodic table invented without them is extraordinary evidence in favor of the validity of Mendeléev's concept.

In order to keep argon (Ar) in its right place in the inert gas family it has to be put ahead of potassium (K) even though that inverts the molecular weight order. Again, this turned out to be right, as also in the case of tellurium (Te) and iodine (I).

Notice also that in Table 1 the five elements with the highest atomic weights were undiscovered in Mendeléev's time. These are polonium (Po), astatine (At), radon (Rn), francium (Fr) and radium (Ra). These were discovered in the 1890s and afterward and are examples of radioactive elements. All are unstable and exist in the earth's crust in vanishingly small quantities. Since these are all at the end of the table, their absence does not affect the remainder.

Then there is fluorine (F) which, strictly speaking, was not known in Mendeléev's time. It is a special case, however. Compounds of fluorine were known and the element is a member of so tightly knit a family that its existence and properties were certain, based on the knowledge of those compounds. It was just that fluorine held on so tightly to other elements that it wasn't till 1886 that chemists could break it away and study it in its elemental form. Actually, it was included in the table from the start (just as the North and South poles could be placed on nineteenth-century globes of the Earth even though no one had yet reached them).

That left the two elements gallium (Ga) and germanium (Ge). These weren't at the end of the table either, in the sense of being in the last column or in the bottom row so that they couldn't be left out without affecting the rest of the table. Their existence, unlike that of fluorine, was completely unsuspected and they left 'holes' in the table.

This means that if you try to list the elements in order of atomic weight and disregard gallium and germanium, you would be forced to place arsenic (As) on the right of aluminum (Al), selenium (Se) on the right of silicon (Si) and so on. This would utterly upset the family and valence arrangements.

151

Mendeléev refused to do that and that was the greatest of his contributions. He put arsenic (As) to the right of phosphorus (P), and selenium (Se) to the right of sulfur (S) where they belonged by every criterion of properties. Since that left two holes to the right of aluminum (Al) and silicon (Si), he calmly decided that they represented two elements that remained yet to be discovered. He called them 'eka-aluminum' and 'eka-silicon' respectively; the 'eka' being the Sanskrit word for 'one.' The missing elements were one place to the right of aluminum and silicon, respectively, in other words.

What's more, Mendeléev predicted the properties of the missing elements in great detail by assuming that gallium (Ga) would have properties midway between aluminum (Al) and indium (In) and that germanium (Ge) would have properties midway between silicon (Si) and tin (Sn).

On the whole, chemists smiled indulgently at the mad Russian, but in 1875 gallium was discovered and in 1886 germanium was discovered and Mendeléev's predictions checked out in every respect. Chemists stopped laughing.

Does this mean that the periodic table as so far described is perfect?

Alas, no. The version of the periodic table present in Table 1 contains just forty-four elements, yet there are many more than that number. Such well-known elements as gold, silver, copper, iron, platinum, manganese and tungsten (all perfectly familiar in Mendeléev's time) find no place in the periodic table in the form presented in Table 1.

Must the periodic table be thrown out then, or can space be found for the additional elements?

Well, notice the three places I have marked with a # mark. Between calcium (Ca) and gallium (Ga) there is an atomic weight difference of 29.7; between strontium (Sr) and indium (In) a difference of 27.2; and between barium (Ba) and thallium (Tl) a difference of fully 67.1. These differences are much larger than exist anywhere else in the periodic table. Indeed, if these three intervals are disregarded, then the average difference in atomic weight from element to element in all the rest of the table is only 2.5.

If we accept 2.5 as the average atomic weight difference between adjoining elements throughout the table, then there is room for twelve elements between calcium (Ca) and gallium (Ga), for eleven between strontium (Sr) and indium (In) and for no less than twenty-seven between barium (Ba) and thallium (Tl).

Is this possible?

Yes, it is, if we get it through our heads that the periods in the periodic table need not necessarily be all the same length (as some of the early speculators had assumed) but could grow longer as one went down the list of elements.

In Mendeléev's time, for instance, the first period had only one member, hydrogen (H), while Periods 2 and 3 had seven members apiece. A generation later, when the inert gases were discovered, it turned out that the first period contained two elements and the second and third periods eight apiece. (There has been no change since.) Why, then, could not the later periods jump to twenty or even to thirty or more?

Indeed, in Mendeléev's time, no less than nine elements were known with atomic weights between those of calcium (Ca) and gallium (Ga), elements that would therefore serve to bridge that large atomic weight gap. Similarly, there were nine elements that would contribute toward a bridging of the gap between strontium (Sr) and indium (In).

The trouble was that valence was no longer as paramount and clear a phenomenon among those elements within the gap as in the elements of Table 1. The elements bridged the gap from one with a clear valence of 2 to one with a clear valence of 3; from calcium (Ca) to gallium (Ga) in the first case and from strontium (Sr) to indium (In) in the second, and because they represented a kind of transition from 2 to 3, they can be called the 'transition elements.' For purposes of this chapter, I will call the elements of Table 1 the 'valence elements.'

We can be guided in arranging the transition elements partly by molecular weight, partly by less clear-cut valence properties, and partly by other chemical properties. In doing so, we can take the eighteen known elements of the first two gaps (as of 1869) and arrange them as in Table 2.

153

There is no real doubt about the arrangement. It is clear, for instance, that silver (Ag) must be to the right of copper (Cu) and that cadmium (Cd) must be to the right of zinc (Zn) out of overwhelmingly convincing chemical considerations. So with the others. Only with the arrangement indicated do properties

TABLE 2 – THE TRANSITION ELEMENTS

		Period 4	Period 5
Valence 2		Ca (40.1)	Sr (87.6)
	a		Y (88.9)
	b	Ti (47.9)	Zr (91.2)
	c	V (50.9)	Nb (92.9)
	d	Cr (52.0)	Mo (95.9)
	e	Mn (54.9)	
Transition Elements	f	Fe (55.8)	Ru (101.7)
	g	Co (58.9)	Rh (102.9)
	h	Ni (58.7)	Pd (106.4)
	i	Cu (63.6)	Ag (107.9)
	j	Zn (65.4)	Cd (112.4)
3		Ga (69.7)	In (114.8)

match left and right – and they do so in the proper order of atomic weight, except for cobalt (Co) and nickel (Ni). There, in order to preserve the chemical verities, the atomic weight order must be inverted, but the atomic weights are so close to each other that the inversion is a rather venal fault. (This is the third – and last – case of an inverted atomic weight order in the periodic table.)

With the eighteen transition elements of Periods 4 and 5 arranged as in Table 2, two holes are found to exist. One is to the left of yttrium (Y), and the other to the right of manganese

154

(Mn). Mendeléev chose the hole to the left of yttrium (Y) as still a third place where he could predict the existence of an undiscovered element, complete with all its properties. (He called it 'eka-boron' because in his first version of the table, he placed the hole to the right of boron [B].)

He was borne out in 1879, when scandium was discovered (see Chapter 11). Its symbol is Sc and its atomic weight is 45.0, fitting it snugly between calcium (Ca) and titanium (Ti).

The hole to the right of manganese (Mn) was not so easily filled. Indeed the element that fits there was not discovered till 1937. It was named technetium (Tc, atomic weight 99).

The atomic weight gaps between the transition elements (assuming ten apiece in Periods 4 and 5, counting the two holes) were about right. The atomic weight differences averaged out to 2.6 as compared with 2.5 for the valence elements.

Could one be sure, in the absence of overriding valence considerations, however, that there might not be eleven elements in each of the two series of transition elements, or even twelve? Suppose, for instance, there were an element missing between c and d in each of the two series. If an element were missing between c and d in only one of the series, we could see the hole it made from the presence of the equivalent element in the other series (as in the case of the hole to the left of yttrium, for instance), but if *both* series were lacking at the same point, we couldn't tell. (That was the case of the inert gases, for when the entire series was unknown, its existence was unsuspected. As soon as one of them was discovered, the others appeared as holes, were searched for and found.)

An argument in favor of ten as the correct number for the transition elements arises from the fact that the total number of elements in Period 4 and Period 5, valence plus transition, is 18, and this introduces an interesting regularity. That is, the total number of elements in Period 1 is $2 \times 1^2 = 2$; the total number in Period 2 and in Period 3 is $2 \times 2^2 = 8$; and the total number in Period 4 and in Period 5 is $2 \times 3^2 = 18$.

This is pretty and to a person with my bent for numbers, for instance, it is even convincing, but actually, what have the elements to do with this neat arrangement? There was no theory at any time during the nineteenth century that would account for such a relationship and it might well be nothing more than a misleading coincidence.

So chemists couldn't be *sure* and the periodic table, though a valuable guide, remained rickety.

Next, what about the third series of transition elements, the ones which must bridge the particularly large atomic weight gap between barium (Ba) and thallium (Tl)? In that gap, in Mendeléev's time, there were eleven elements known. If we try to make them match the other two series of transition elements in the *a* to *j* scheme, we end up with Table 3.

TABLE 3 – MORE TRANSITION ELEMENTS

	Valence	Period 6
	2	Ba (137.3)
a		La (138.9)
b		
c		Ta (180.9)
d		W (183.9)
e		
f		Os (190.2)
g		Ir (192.2)
h		Pt (195.1)
i		Au (197.0)
j		Hg (200.6)
	3	Tl (204.4)

Transition Elements

The elements shown in Table 3 match up indubitably with those in Table 2. Thus gold (Au) is clearly in position *i* to the right of copper (Cu) and silver (Ag), and the rest of the elements shown belong in their places with equal clarity.

This leaves two holes, though. In position *b* there ought to be an element to the right of zirconium (Zr) and in 1923 that element was indeed discovered. It was named hafnium. (Hf, atomic weight 178.5) and was discovered in zirconium ores. It fit the place perfectly; too perfectly. It took so long to discover hafnium not because that element was excessively rare, but because it was so like zirconium in all its properties that it was difficult to separate from its fiftyfold more common twin sister.

The hole in position e was filled in 1925 with the discovery of rhenium (Re, molecular weight 186.2).

There was no element discovered in the third transition series which indicated the existence of unsuspected holes in the first or second transition series. That was a point in favor of supposing the ten elements in each of those first two series to be all there were.

But even allowing for the discovery of hafnium (Hf), you will notice that there is a sizable atomic weight gap between that and lanthanum (La), a gap of 39.6. This gap exists between a and b of Period 6 and there is no such gap at all in the corresponding position of Period 4 or Period 5. There is room for a number of elements in this gap.

Yet I said that there were eleven elements known in Mendeléev's time with atomic weight lying between those of barium (Ba) and thallium (Tl). Table 3 only accounted for eight of them. What of the other three?

Those other three have atomic weights that do indeed fall in the new gap between lanthanum (La) and hafnium (Hf) and they are cerium (Ce), erbium (Er) and terbium (Tb).

These are three of the rare earth metals which I discussed in the previous chapter. Two others were known at the time: lanthanum (La) and yttrium (Y) and still another was shortly discovered, scandium (Sc). Scandium, lanthanum and yttrium, however, all fit into position a of Periods 4, 5, and 6 respectively and are ordinary transition elements. It is only cerium, erbium and terbium that are to be placed in this special gap in Period 6. By 1907 ten more rare earth elements were located with atomic weights that placed them in this special gap. The list of thirteen are present in Table 4.

How many more might there be?

Suppose we go back to the number game I introduced a little

while ago. The same system that explains the numbers 2, 8, 8, 18, 18 for the first five periods would make the total number of elements in the sixth period $2 \times 4^2 = 32$. Since the valence elements and transition elements together in the sixth period make up 18, that would leave 14 rare earth elements to bridge the gap and add up to 32.

TABLE 4 – RARE EARTH ELEMENTS

Period 6

a Lanthanum (La)
    (138.9)

Cerium (Ce)	140.1
Praseodymium (Pr)	140.9
Neodymium (Nd)	144.2
Samarium (Sm)	150.4
Europium (Eu)	152.0
Gadolinium (Gd)	157.3
Terbium (Tb)	158.9
Dysprosium (Dy)	162.5
Holmium (Ho)	164.9
Erbium (Er)	167.3
Thulium (Tm)	168.9
Ytterbium (Yb)	173.0
Lutetium (Lu)	175.0

b Hafnium (Hf)
    (178.5)

We have 13; where would we find the 14th?

Between neodymium (Nd) and samarium (Sm) there is an atomic weight difference of 6.2, over twice the normal value. It might be there. However, the difference between europium (Eu) and gadolinium (Gd) is 5.3 and that between thulium (Tm) and ytterbium (Yb) is 4.1. Perhaps there are three missing rare earth elements, one in each place, or, who knows, even more. We can't after all bind ourselves too firmly to a pretty numerical relationship in the absence of physical evidence that would explain its existence.

In short, forty years after Mendeléev had presented the periodic table of the elements, it remained incomplete. Despite the enormous triumphs it had achieved and the neat manner in which it had solved almost every problem it had faced, chemists could not be sure it would remain an adequate guide under all conditions. In particular, they could not be sure it could account, adequately, for the rare earth elements.

It was for this reason, more than any other, that chemists anxiously combed through the rare earth minerals to see how many new elements they could indubitably identify. By doing so they might bring the entire periodic table down with a crash.

They didn't. Instead, in 1914, the periodic table was placed on a firm, logical footing at last, and that happened in an utterly unexpected way through a line of research that seemed to have nothing to do with chemistry. We'll track that down in the next chapter.

13

The Nobel Prize That Wasn't

Some time ago, I gave a lecture at a nearby university and the evening began with a dinner which deserving students were allowed to attend. Naturally, the attendees were science fiction fans who thought it would be great to meet me, and that suited me fine because I think it's great to meet people who think it's great to meet me.

One of the students was a buxom eighteen-year-old coed and I found that delightful, because many years ago I took a liking to buxom eighteen-year-old coeds and I've never entirely outgrown that feeling. She sat next to me at the dinner and I was at my genial and witty best, simply oozing gallantry and charm. Somewhere around the dessert, though, I paused for breath and, in the silence, the sound of the conversation elsewhere along the table welled up about us.

We both stopped to listen. It was the other collegiates talking; all of them earnest young men and women, deeply involved in the burning issues of the day. To be sure, I was about to give a talk on the burning issues of the day, but, even so, listening to the others made me feel a little ashamed that I had burdened my companion of the meal with nothing more than nonsense. And just as I was beginning to launch into some deep philosophy, she said to me, 'Everyone is so serious here. Ever since I came to college I've met only serious people.'

She paused to think and then said, with every sign of absolute sincerity, 'In fact, in all the time I've been here, you're the first eighteen-year-old I've met.'

So I kissed her.

But you know, however youthful I feel and act in consequence of my temperament, my way of life and my constant association with college students, I am nevertheless over eighteen. My enemies might even say I was far, *far* beyond eighteen and they would be right.

Still, there's no way of avoiding the advance of years except by dying, and there's no great fun in that, as I will show you in the case of one young man who will be under discussion in this chapter —

In the previous chapter, I talked about the periodic table, and how even after nearly half a century of steady triumphs, it still lacked a firm foundation in the second decade of the twentieth century. It received that foundation, finally, through a twenty-year set of development that began with something seen out of the corner of the eye.

The year of that beginning was 1895, the place was in the laboratory of Wilhelm Konrad Roentgen, head of the physics department at the University of Würzburg in Bavaria. Roentgen was investigating cathode rays – the big glamor object of physics in those days. An electric current forced through a good enough vacuum emerged as a stream of what turned out to be particles much smaller than atoms (subatomic particles), which received the name of 'electrons.'

These streams of electrons had a host of fascinating properties. For one thing, they produced luminescence when they struck certain chemicals. The luminescence wasn't very bright so, in order to study it more easily, Roentgen darkened the room and encased the cathode-ray tube in thin black cardboard.

On November 5, 1895, then, he turned on his cathode-ray tube and prepared to peer close inside the box and proceed with his experiments. Before he could do so, a sparkle of light in the darkness caught his eye. He looked up and there, to one side of the tube was a piece of paper coated with barium platinocyanide, one of the chemicals that glowed when struck by the fleeting electrons.

What puzzled Roentgen was that the barium platinocyanide didn't happen to be in the path of the electrons. If the paper had been *inside* the cardboard box at the proper end of the cathode-ray tube, why all right. But the glowing paper was to one side of the tube and even if one supposed that some of the electrons were leaking sideways, there was no way they could get through the cardboard.

Perhaps the glow was caused by something else altogether and had nothing to do with the cathode-ray tube. Roentgen shut off the electric current, the cathode-ray tube went dead – and the coated paper stopped glowing. He turned the electric current on and off and the coated paper glowed and ceased glowing in exact rhythm. He took the paper into the next room and it glowed (more faintly) only when the cathode-ray tube went into operation.

Roentgen could only come to one conclusion. The cathode-ray tube was producing some mysterious radiation that was extraordinarily penetrating; that could go through cardboard and even walls. He hadn't the faintest notion of what that radiation might be so he named it with the symbol of the unknown. He called it 'X rays' and it has kept that name ever since.

Roentgen experimented furiously and then, after a phenomenally short interval, managed to publish the first paper on the subject on December 28, 1895, reporting all the basic properties of the new radiation. On January 23, 1896, he gave his first public lecture on the phenomenon. He produced X rays before the excited audience, showed that they would fog a photographic plate and that they would penetrate matter – some types of matter more easily than others.

X rays would penetrate the soft tissues, for instance, more easily than bone. If a hand were placed on a photographic plate and exposed to X rays, the bones would block so much of the X rays that the portion of the plate under them would remain relatively unfogged. The bones would appear white on black. An aged Swiss physiologist, Rudolf Albert von Kölliker, volunteered, and an X-ray photograph of his hand was taken.

No physical discovery was ever applied to medical science so quickly. The thought that the interior of intact, living organisms could be seen caused intense excitement and only four days after the news of X rays reached the United States, the new radiation was successfully used to locate a bullet in a man's leg. Within a year of Roentgen's discovery, a thousand papers on X rays were published and in 1901, when the Nobel prizes were first set up, the very first to be awarded in physics went to Roentgen.

(Laymen went wild, too. Panicky members of the New Jersey legislature tried to push through a law preventing the use of X rays in opera glasses for the sake of maidenly modesty – which was about par for legislative understanding of science.)

It was clear that the radiation couldn't appear out of nowhere. The speeding electrons making up the cathode rays struck the glass of the tube and were stopped more or less suddenly. The kinetic energy of those speeding electrons had to appear in another form, and they did so as X rays, which were energetic enough to smash through considerable thicknesses of matter.

If this happened when electrons struck glass what would happen when they struck something which was denser than glass and could stop them more effectively? The greater deceleration ought to produce more energetic X rays than those Roentgen had first observed. Pieces of metal were therefore sealed into the cathode-ray tubes in places where they would be struck by the electrons. The expected happened. Larger floods of more energetic X rays were produced.

The X rays produced by the collision of electrons with metals were studied with particular care in 1911 by the English physicist Charles Glover Barkla. Physicists had not yet worked up appropriate techniques for measuring the properties of X rays with real delicacy but one could at least tell that one particular sheaf of X rays might penetrate a greater thickness of matter than another sheaf would and that the first therefore contained more energy.

Barkla found that for a given metal, X rays were produced in sharply different energy ranges, judging by their penetrating quality. There would be what he called the K series, the L series, the M series and so on, in order of decreasing penetrability and, therefore, decreasing energy content. The energy range was discontinuous. There were no X rays to speak of at energy levels intermediate between the K and the L or between the L and the M and so on.

What's more, each different metal produced a set of X rays with energies characteristic of itself. If one focused on one particular series, the L series, for instance, these would increase in energy the higher the atomic weight of the metal that was stopping the electrons.

Since the X-ray energy levels were characteristic of the metal used to stop the electrons, Barkla called them 'characteristic X rays.'

The *x* of X rays remained appropriate for seventeen years after Roentgen's initial discovery.

Were X rays composed of particles like electrons, but much more energetic? Or were X rays made up of bundles of electromagnetic waves like those of ordinary light, but much more energetic?

If X rays consisted of waves, they would be bent in their course by a diffraction grating, one in which there were numerous fine, opaque lines, parallel to each other, on an otherwise transparent screen. The trouble was that the lines in such gratings have to be separated by small distances. The shorter the wavelengths of the radiation being studied, the more closely spaced the diffraction lines must be.

One could rule, by mechanical means, lines fine enough and closely spaced enough to diffract ordinary light waves, but if X rays were like light but much more energetic, their waves would have to be much smaller than those of light. Lines simply could not be ruled close enough to handle X rays.

It occurred to a German physicist, Max Theodor Felix von Laue, that one did not have to depend on man-made lines. Crystals consisted of atoms arranged in great regularity. Within the crystal there would be sheets of atoms of one particular kind oriented along one particular plane. There would be successive sheets of these atoms separated by just the distances one would need for diffracting X rays. A crystal, in other words, was a diffraction grating designed by Nature for use in the study of X rays (if one wanted to be romantic about it).

Well, then, if X rays were sent through a crystal and if they were diffracted in a way one could predict from theory, *assuming* the X rays were light-like waves, then the X rays very likely *were* light-like waves.

In 1912, Von Laue and his associates sent a beam of X rays

through a crystal of zinc sulfide and it *was* diffracted just so. The X rays were electromagnetic radiation then, like light but far more energetic. Now X rays were no longer *x*, but they kept the name anyway.

Scientists could go further. The distance between sheets of atoms in the crystal could be worked out from data not involving X rays. From this, one could calculate how much diffraction different wavelengths ought to yield. By passing X rays through a given crystal of a pure substance, then, and measuring the amount of diffraction (something that was reasonably easy to do) the wavelength of a particular set of X rays could be determined with surprising precision.

A young Australian student of physics at Cambridge, William Lawrence Bragg, hearing of Von Laue's experiment, saw the point at once. His father, who was teaching physics at the University of Leeds, saw the same point. Together, father and son began measuring X-ray wavelengths at a great rate and perfected the technique.

And this brings me to the hero of this chapter, the English physicist Henry Gwyn-Jeffreys Moseley, son of a professor of anatomy who died when Henry was only four.

Moseley was simply a streak of brilliance. He won scholarships to both Eton and Oxford and in 1910, when he was twenty-three years old, he joined the group of young men who were working under the New Zealand-born Ernest Rutherford at Victoria University in Manchester, and stayed with him for two years.

Rutherford was himself one of the great experimenters of all times and had won the Nobel Prize in 1908. (He won it in chemistry, because his physical discoveries had such exciting significance for the science of chemistry – rather to his disgust, for like any good physicist he tended to look down on chemists.)

What's more, seven of those who worked for him at one time or another went on to win Nobel prizes of their own eventually. Yet there is room to argue that of all those who worked for Rutherford, none was more brilliant than Moseley.

It occurred to Moseley to combine the work of the Braggs

and of Barkla. Instead of differentiating among the various characteristic X rays associated with different metals by Barkla's rather crude criterion of penetrability, he would send them through crystals, *à la* the Braggs, and measure their wavelengths with precision.

This he did in 1912 (by which time he had shifted to Oxford and to independent research) for the metals calcium, titanium, vanadium, chromium, manganese, iron, cobalt, nickel and copper. These elements make up, in that order, a solid stretch across the periodic table – except that between calcium and titanium there should be scandium and Moseley had no scandium available with which to work.

Moseley found a particular series of the characteristic X rays associated with each metal decreased in wavelength (and therefore increased in energy) as one went up the periodic table and did so in a very regular way. In fact, if you took the square root of the wavelength, the relationship was a straight line.

This was extraordinarily important because the atomic weights which, until then, had been the chief way of judging the order of the elements in the periodic table showed no such great regularity. The atomic weights of the elements studied by Moseley were (to one decimal place): 40.1, 47.9, 50.9, 52.0, 54.9, 55.8, 58.9, 58.7, and 63.5. The atomic weight of scandium which Moseley did not have available was 45.0. The atomic weight intervals are, therefore, 4.9, 2.9, 3.0, 1.1, 2.9, 0.9, 3.1, —0.2, 4.8.

These irregular intervals simply could not compare with the absolute regularity of the X-ray wavelengths. What's more, in the periodic table there were occasional places where elements were out of order if the atomic weights were used as criteria. Thus, from their chemical properties, it was certain that nickel came after cobalt in the table, even though nickel's atomic weight was slightly lower than that of cobalt. This *never* happened with X-ray wavelength. By that criteria, nickel had characteristic X rays of greater energy than cobalt and *ought* to come after cobalt.

The conclusion Moseley was forced to come to was that the atomic weight of an element was not a fundamental characteristic and did not entirely, in and of itself, account for why a particular element was a particular element. The X-ray wavelengths, on the other hand, represented something that *was* a fundamental characteristic of the elements.

Moseley was even able to point out what that something was.

Just one year before, Moseley's old boss, Rutherford, had conducted a series of elegant experiments that had demonstrated the basic principles of atomic structure. The atom was not the featureless, ultimate particle it had been thought to be all through the nineteenth century. Instead, it had a complex internal makeup.

Almost all the atomic mass was concentrated in the very center of its structure, in an 'atomic nucleus' that took up only a quadrillionth of the volume of the atom. All about it, filling the rest of the atom, were electrons which were mere froth, for one electron had a mass only 1/1837 that of even the lightest atom (that of hydrogen).

Each electron had a unit negative charge which was absolutely identical in size in all electrons (as far as anyone knew then, or, for that matter, now). The electron charge is usually represented as -1.

The atom as a whole, however, was electrically uncharged. It followed therefore that the central atomic nucleus must have a balancing positive charge.

Suppose, then, that each different element was made up of atoms containing a characteristic number of electrons. The central nuclei of these atoms must contain that same characteristic and balancing number of positive unit charges. If an element had atoms containing only one electron, its nucleus would have a charge of $+1$. An atom with two electrons would have a nucleus with a charge of $+2$. One with three electrons, a nucleus with a charge of $+3$ and so on.

Electrons in varying numbers can, however, be stripped from or added to particular atoms, leaving those atoms with a net positive or negative charge respectively. This means that the electron number is not really fundamentally crucial to the nature of the atom. The atomic nucleus, hidden far within the center of the atom, could not be manipulated by ordinary chemical methods, however. It remained a constant factor and it was therefore *the* characteristic property of an element.

In Moseley's time, nobody knew the details of the structure of the atomic nucleus, of course, but that was not yet necessary. The size of the positive charge on the nucleus was enough.

It was easy to argue, for instance, that the speeding electrons of the cathode rays would be decelerated more effectively as the charge content of the atom they struck increased. The energy of the X rays produced would increase in some regular fashion with the increase in charge content; and if the charge content increased very regularly by unit charges, then so would the energy content of the X rays.

Moseley suggested that each element be represented by a number that would express two different things – (1) the number of unit positive charges on the nuclei of its atoms and (2) its position in the periodic table.

Thus, hydrogen, as the first element in the table, would be represented by the number 1 and, it was to be hoped, would have 1 unit positive charge on its atomic nucleus (this turned out to be correct). Helium would be 2, this representing the fact that it was the second element in the periodic table and had two unit positive charges on the nuclei of its atoms. And so on, all the way to uranium, the last element then known in the periodic table, which would, from the data gathered then and since, have ninety-two unit charges on its atomic nuclei and therefore be represented by the number 92.

Moseley suggested that these numbers be called 'atomic numbers' and that suggestion was adopted.

Moseley published his findings in 1913 and they made an enormous splash at once. In Paris, Georges Urbain thought he would test Moseley. He had spent many years carefully and painstakingly separating rare earth minerals and he prepared a mixture of several oxides which he felt no one but an expert could analyze, and that only after long and tedious fractionations. He brought it to Oxford and there Moseley bounced electrons off the mixture, measured the wavelength of the X rays produced and in hardly any time at all announced the mixture to contain erbium, thulium, ytterbium, and lutetium – and he was right.

Urbain was astonished, as much by Moseley's youth (he was still only twenty-six) as by the power of his discovery. He went back to Paris, preaching the atomic number concept with fervor.

Now at last the periodic table was on a firm foundation. When the X-ray wavelengths differed by a certain known minimum amount, then two elements were adjacent and had nuclear charges that differed by a single unit. There could be *no new elements located between them.*

This meant that from hydrogen to uranium inclusive, there were exactly ninety-two conceivable elements, no more and no less. And in the half century since Moseley's discovery no unexpected elements in the hydrogen-uranium range have showed up between two elements predicted adjacent by X-ray data. To be sure, new elements were located beyond uranium, with atomic numbers of 93, 94, and so on, up to (at present writing) 104 and *possibly* 105, but that is a different story.

Furthermore, if the X-ray wavelengths of two elements differed by twice the expected interval then there *was* an element in between, exactly *one* element. If no such element was known, then it remained to be discovered, that was all.

At the time the atomic number concept was advanced, eighty-five elements were known in the range from hydrogen to uranium. Since there was room for ninety-two elements, it meant that there still remained exactly seven undiscovered elements. What's more their atomic numbers turned out to be 43, 61, 72, 75, 85, 87, and 91.

This solved the problem posed in the previous chapter concerning the total number of rare earths. It turned out there was only one rare earth not yet discovered and it was located in number 61, between neodymium (60) and samarium (62). It took over thirty years to discover the missing seven elements and as it happened, the very last to be discovered was the rare earth, 61. It was discovered in 1948 and named promethium. (By that time, though, elements beyond uranium were being discovered.)

Thanks to Moseley's atomic number concept, the foundation of the periodic table was made firm as rock. Every discovery since then has served only to strengthen both the atomic number and the periodic table.

Clearly, Moseley deserved the Nobel Prize in either physics

or chemistry (toss a coin and take your pick, and I could argue that he deserved one of each), and it was just as certain as anything could be in such matters that he was going to get it.

In 1914, the physics prize went to Von Laue and in 1915 to the father–son combination of the Braggs. In both cases the work on X rays had served as preliminaries to the culminating work of Moseley. In 1916, then, Moseley would have *had* to get it; there was no way of avoiding it.

I'm sorry; there *was* a way of avoiding it.

In 1914, World War I broke out and Moseley enlisted at once as a lieutenant in the Royal Engineers. That was his choice and he is to be respected for his patriotism. Still, just because an individual is patriotic and wishes to risk a life that is not entirely his own to throw away doesn't mean that the decision makers of a government have to go along with it.

In other words, Moseley might have volunteered a thousand times and yet the government had no business sending him to the front. Rutherford understood this and tried to have Moseley assigned to scientific labors since it was obvious that he could be far more valuable to the nation and the war effort in the laboratory than in the field. By World War II, this was understood and Moseley would have been protected as a rare and valuable war resource.

No such thing was to be expected in the monumental stupidity that was called World War I.

In the spring of 1915, the British got the idea of landing at Gallipoli in western Turkey to seize control of the narrow straits linking the Mediterranean and Black seas. Forcing a passage through, they could open a supply route to the tottering Russian armies, which combined enormous individual bravery with equally enormous administrative ineptitude. Strategically, the concept was a good one, but tactically it was handled with incredible folly. Even in a war so consistently idiotic, the Gallipoli campaign manages to shine as an archetype of everything that should not be done.

By January 1916, it was all over. The British had thrown in half a million men and gotten nowhere. Half of them were casualties.

In the course of this miserable campaign, Moseley was

tapped. On June 13, 1915, he embarked for Gallipoli. On August 10, 1915, while he was telephoning an order, a Turkish bullet found its mark. He was shot through the head and killed at once. He had not yet reached his twenty-eighth birthday and, in my opinion, his death was the most expensive individual loss to the human race generally, among all the millions who died in that war.

When the time for the 1916 Nobel Prize in physics came about, there was no award. It is easy to explain that by saying that the war was on, but there had been an award in 1915 and there was to be one in 1917. The 1917 one was to Barkla, still another man whose work was only preliminary to the great breakthrough of Moseley's.

Call me sentimental, but I see no reason why the colossal stupidity of the human race should force the indefinite perpetration of a disgraceful injustice. It is not too late, even now, for the community of science to fill that gap and to state that the 1916 Nobel Prize in physics (that wasn't) belongs to Moseley, and that he ought to appear in every list of Nobel laureates published.

We don't owe it to him; I'm not *that* sentimental. He is beyond either debt or repayment. We owe it to the good name of science.

D
Sociology

14
The Fateful Lightning

In the last five years or so, I have turned to writing history. I don't mean the history of science (I've been doing that for a long time); I mean 'straight' history. As of now, I have published seven history books, with more to come.

This is valuable to me in a number of ways. It keeps my fingers nimbly stroking the typewriter keys and it keeps my mind exercised in new and refreshing directions. And, both least and most important, it inveigles me into new games.

No one who reads these essays can help knowing that I love to play with numbers — Well, I discovered I love to play with turning points, too. There's the excitement of tracing down an event and saying: 'At this point, at this exact point, man's history forked and man moved irrevocably into this path rather than the other.'

To be sure, I'm somewhat of a fatalist and believe that 'man's history' is the product of rather massive forces that will not be denied; that if a certain turning is prevented at this point, it will come about at another point eventually. Yet even so, it remains interesting to find the point where the turning *was* made.

Of course, the most fun of all is to find a brand-new turning point; one which has never (to one's knowledge) been pointed out. My own chance at finding a new turning point is made somewhat better than it might be, in my opinion, by my advantage of being equally at home in history and in science.

By and large, historians tend to be weak in science and they find their turning points in political and military events for the most part. Such watershed years of history as 1453, 1492, 1517, 1607, 1789, 1815 and 1917 have nothing directly to do with science. Scientists, on the other hand, tend to think of science in terms rather divorced from society and such turning-point years as 1543, 1687, 1774, 1803, 1859, 1895, 1900, and

1905 tend to have no immediate and direct connection with society.*

To me, however, a turning point of the first magnitude, one that is *equally* important both to science and to society, took place in 1752, and no one, to my knowledge, has ever made an issue of it. So, Gentle Reader, *I* will—

As far as our records go back and, presumably, much further, men have turned to experts for protection against the vagaries of nature.

That protection they surely needed, for men have been subjected to seasons of bad hunting when they were hunters, and to seasons of sparse rainfall when they were farmers. They have fallen prey to mysterious toothaches and intestinal gripings; they have sickened and died; they have perished in storms and wars; they have fallen prey to mischance and accident.

All the Universe seemed to conspire against poor, shivering man, and yet it was, in a way, his transcendent triumph that he felt there must be some way in which the tables could be turned. If only he had the right formula, the right mystic sign, the right lucky object, the right way of threatening or pleading – why, then, game would be plentiful, rain would be adequate, mischance would not befall, and life would be beautiful.

If he didn't believe that, then he lived in a Universe that was unrelievedly capricious and hostile, and few men, from the Neanderthal who buried his dead with the proper ceremony, to Albert Einstein who refused to believe that God would play dice with the Universe, were willing to live in such a world.

Much of human energies in prehistory, then, and in most of historical times, too, went into the working out of the proper ritual for control of the Universe and into the effort of establishing rigid adherence to that ritual. The tribal elder, the patriarch, the shaman, the medicine man, the wizard, the magician, the seer, the priest, those who were wise because they were old, or wise because they had entry into secret teachings, or wise

* You're welcome to join the fun of turning-pointing by trying to figure out what happened in these years, without looking them up, but you don't have to. The details are not relevant to the remainder of the chapter.

simply because they had the capacity to foam at the mouth and go into a trance, were in charge of the rituals and it was to them that men turned for protection.

In fact, much of this remains. Verbal formulas, uttered by specialists, are relied on to bring good luck to a fishing fleet, members of which would be uneasy about leaving port without it. If we think this is but a vagary of uneducated fishermen, I might point out that the Congress of the United States would feel most uneasy about beginning its deliberations without a chaplain mimicking biblical English in an attempt to rain down good judgment upon them from on high – a device that seems very rarely to have done the Congress much good.

It is not long since it was common to sprinkle fields with holy water to keep off the locusts, to ring church bells to keep off earthquakes and counter the deadly effects of comets, to use united supplications according to agreed-upon wording to bring on needed rain. In short, we have not really abandoned the attempt to control the Universe by magic.

The point is that well into the eighteenth century, there was no other way to find security. Either the Universe was controlled by magic (whether through spells or through prayer) or it couldn't be controlled at all.

It might *seem* as though there *was* an alternative. What about science? By the mid-eighteenth century, the 'scientific revolution' was two centuries old and had already reached its climax with Isaac Newton, three quarters of a century before. Western Europe, and France in particular, was in the very glory of the 'Age of Reason.'

And yet science was not an alternative.

In fact, science in the mid-eighteenth century still meant nothing to men generally. There was a tiny handful of scholars and dilettantes who were interested in the new science as an intellectual game suitable for gentlemen of high IQ, but that was all. Science was a thoroughly abstract matter that did not (and, indeed, according to many scientists in a tradition that dated back to the ancient Greeks, *should* not) involve practical matters.

Copernicus might argue that the Earth went around the

176

Sun, rather than vice versa; Galileo might get into serious trouble over the matter; Newton might work out the tremendous mechanical structure that explained the motions of the heavenly bodies – yet how did any of that affect the farmer, the fisherman or the artisan?

To be sure there were technological advances prior to the mid-eighteenth century that did affect the ordinary man; sometimes even very deeply; but those advances seemed to have nothing to do with science. Inventions such as the catapult, the mariner's compass, the horseshoe, gunpowder, printing were all revolutionary, but they were the product of ingenious thinking that had nothing to do with the rarefied cerebrations of the scientist (who, in the eighteenth century, was called a natural philosopher for the term 'scientist' had not yet been invented).

In short, as late as the mid-eighteenth century, the general population not only did not consider science as an alternative to superstition, they never dreamed that science could have any application at all to ordinary life.

It was in 1752, exactly, that that began to change; and it was in connection with the lightning that the change began.

Of all the fatal manifestations of nature, the most personal one, the one which is most clearly an overwhelming attack of a divine being against an individual man, is the lightning bolt.

War, disease and famine are all wholesale forms of destruction. Even if, to the true believers, these misfortunes are all the punishment of sin, they are at least punishment on a mass scale. Not you alone, but all your friends and neighbors suffer the ravages of a conquering army, the agony of the Black Death, the famishing that follows drought-killed grainfields. Your sin is drowned and therefore diminished in the mighty sin of the village, the region, the nation.

The man who is struck by lightning, however, is a personal sinner, for his neighbors are spared and are not even singed. The victim is selected, singled out. He is even more a visible mark of a god's displeasure than the man who dies of a sudden apoplectic stroke. In the latter case, the cause is invisible and may be anything, but in the former there can be no doubt. The divine displeasure is blazoned forth and there is thus a kind of superlative disgrace to the lightning stroke that goes beyond death and lends an added dimension of shame and horror to the thought of being its victim.

Naturally, lightning is closely connected with the divine in our best-known myths. To the Greeks, it was Zeus who hurled the lightning and to the Norse, the lightning was Thor's hammer. If you care to turn to the 18th Psalm (verse 14 in particular) you will find that the biblical God also hurls lightning. Or as Julia Ward Howe says in her 'Battle Hymn of the Republic' – 'He hath loosed the fateful lightning of His terrible, swift sword.'

And yet, if the lightning stroke were obviously the wrathful weapon of a supernatural being, there were some difficult-to-explain consequences.

As it happens, high objects are more frequently struck by lightning than low objects are. As it also happens, the highest man-made objects in the small European town of early modern times were the steeples of the village church. It followed, embarrassingly enough, that the most frequent target of the lightning bolt, then, was the church itself.

I have read that over a thirty-three-year period in eighteenth-century Germany, no less than four hundred church towers were damaged by lightning. What's more, since church bells were often rung during thunderstorms in an attempt to avert the wrath of the Lord, the bell ringers were in unusual danger and in that same thirty-three-year period, 120 of them were killed.

Yet none of this seemed to shake the preconceived notion that connected lightning with sin and punishment. Until science took a hand.

In the mid-eighteenth century, scientists were fascinated by the Leyden jar. Without going into detail, this was a device which enabled one to build up a sizable electric charge; one which, on discharge, could sometimes knock a man down. The charge on a Leyden jar could be built up to the point where it might discharge across a small air gap and when that happened, there was a brief spark and a distinct crackling sound.

It must have occurred to a number of scholars that the discharge of a Leyden jar seemed to involve a tiny lightning bolt with an accompanying pygmyish roll of thunder. Or, in reverse, it must have occurred to a number of them that in a thunderstorm, earth and sky played the role of a gigantic Leyden jar and that the massive lightning stroke and the rolling thunder were but the spark and crackle on a huge scale.

But thinking it and demonstrating it were two different things. The man who demonstrated it was our own Benjamin Franklin – the 'Renaissance Man' of the American colonies.

In June 1752, Franklin prepared a kite, and to its wooden framework he tied a pointed metal rod. He attached a length of twine to the rod and connected the other end to the cord which held the kite. At the lower end of the cord, he attached an electrical conductor in the shape of an iron key.

The idea was that if an electric charge built up in the clouds, it would be conducted down the pointed rod and the rain-wet cord to the iron key. Franklin was no fool; he recognized that it might also be conducted down to himself. He therefore tied a non-conducting silk thread to the kite cord, and held that silk thread, rather than the kite cord itself. What's more, he remained under a shed so that he and the silk thread would stay dry. He was thus effectively insulated from the lightning.

The strong wind kept the kite aloft and the storm clouds gathered. Eventually, the kite vanished into one of the clouds and Franklin noted that the fibers of the kite cord were standing apart. He was certain that an electric charge was present.

With great courage (and this was the riskiest part of the experiment) Franklin brought his knuckle near the key. A spark leaped across the gap from key to knuckle. Franklin heard the crackle and felt the tingle. It was the same spark, crackle and tingle he had experienced a hundred times with Leyden jars. Franklin then took the next step. He had brought with him an uncharged Leyden jar. He brought it to the key and charged it with electricity from the heavens. When he had done so, he found that that electricity behaved exactly as did ordinary earthly electricity produced by ordinary earthly means.

Franklin had demonstrated that lightning was an electrical

discharge, different from that of the Leyden jar only in being immensely larger.

This meant that the rules that applied to the Leyden jar discharge would also apply to the lightning discharge.

Franklin had noted, for instance, that an electrical discharge took place more readily and quietly through a fine point than through a blunt projection. If a needle were attached to a Leyden jar, the charge leaked quietly through the needle point so readily that the jar could never be made to spark and crackle.

Well, then— If a sharp metal rod were placed at the top of some structure and if that were properly grounded, any electric charge accumulating on the structure during a thunderstorm would be quietly discharged and the chances of its building up to the catastrophic loosing of a lightning bolt were greatly diminished.

Franklin advanced the notion of this 'lightning rod' in the 1753 edition of *Poor Richard's Almanac*. The notion was so simple, the principle so clear, the investment in time and material so minute, the nature of the possible relief so great that lightning rods began to rise over buildings in Philadelphia by the hundreds almost at once, then in New York and Boston, and soon even in Europe.

And it worked! Where the lightning rods rose, the lightning stroke ceased. For the first time in the history of mankind, one of the scourges of the Universe had been beaten, not by magic and spells and prayer, not by an attempt to subvert the laws of nature – but by science, by an understanding of the laws of nature and by intelligent cooperation with them.

What's more, the lightning rod was a device that was important to every man. It was not a scholar's toy; it was a lifesaver for every mechanic's house and for every farmer's barn. It was not a distant theory, it was a down-to-earth fact. Most of all, it was the product not of an ingenious tinkerer, but of a logical working out of scientific observations. It was clearly a product of science.

Naturally, the forces of superstition did not give in without a struggle. For one thing, they made the instant point that since the lightning bolt was God's vengeance, it was the height of impiety to try to ward it off.

This, however, was easy to counter. If the lightning were God's artillery and if it could be countered by a piece of iron, then God's powers were puny indeed and no minister dared imply that they were. Furthermore, the rain was also sent by God and if it was improper to use lightning rods, it was also improper to use umbrellas or, indeed, to use overcoats to ward off God's wintry winds.

The great Lisbon earthquake of 1755 was a temporary source of exultation for the ministers in the churches of Boston. There were not wanting those who pointed out that in his just wrath against the citizens of Boston, God had, with a mighty hand, destroyed the city of Lisbon. This merely succeeded, however, in giving the parishioners a poor notion of the accuracy of the divine aim.

The chief resistance, however, was negative. There was an embarrassed reluctance about putting up lightning rods on churches. It seemed to betray a lack of confidence in God; or worse still, a fullness of confidence in science that would seem to countenance atheism.

But the results of refusing to put up lightning rods proved insupportable. The church steeples remained the highest objects in town and they continued to be hit. It became all too noticeable to all men that the town church, unprotected by lightning rods, was hit, while the town brothel, if protected by lightning rods, was not.

One by one, and most reluctantly, the lightning rods went up even over the churches. It became quite noticeable then, that a particular church whose steeple had been damaged over and over, would stop having any of this kind of trouble once the lightning rod went up.

According to one story I've read, the crowning incident took place in the Italian city of Brescia. The church of San Nazzaro in that city was unprotected by lightning rods but so confident was the population in its sanctity that they stored a hundred tons of gunpowder in its vaults, considering those vaults to be the safest possible place for it.

But then, in 1767, the church was struck by lightning and the gunpowder went up in a gigantic explosion that destroyed one sixth of the city and killed three thousand people.

That was too much. The lightning rod had won and super-stition surrendered. Every lightning rod on a church was evidence of the victory and of the surrender and no one could be so blind as not to see that evidence. It was plain to anyone who would devote any thought to the problem that the proper road to God was not through the self-will of man-made magical formulas but through the humble exploration of the laws governing the Universe.

Although the victory over lightning was a minor one in a way, for the number killed by lightning in the course of a year is minute compared to the number killed by famine, war or disease, it was crucial. From that moment on, the forces of superstition * could fight only rearguard actions and never won a major battle.

Here's one example. In the 1840s, the first really effective anesthetics were introduced and the possibility arose that pain might be abolished as a necessary accompaniment of surgery and that hospitals might cease to be the most exquisitely organized torture chambers in the history of man. In particular, anesthesia might be used to ease the pains of childbirth.

In 1847, a Scottish physician, James Young Simpson, began to use anesthesia for women in labor, and at once the holy men mounted their rostrums and began their denunciations.

From pulpit after pulpit, there thundered forth a reminder of the curse visited upon Eve by God after she had eaten of the fruit of the tree of the knowledge of good and evil. Male ministers, personally safe from the pain and deadly danger of childbearing, intoned: 'Unto the woman he said, I will greatly multiply thy sorrow and thy conception; in sorrow thou shalt bring forth children. . . .' (Genesis 3:16).

The usual story is that those apostles of mothers' anguish, these men who worshipped a God whom they viewed as willing to see hundreds of millions of agonized childbirths in each

---

* I am saying superstition, *not* religion. The ethical and moral side of religion is not involved in the fight against the lightning rod or against any other scientific finding. Only traditional superstitious beliefs are in the fight and it may well be argued that these are even more harmful to real religion than they are to science and rationality.

generation, when the means were at hand to ease the pain, were defeated by Simpson himself through a counterquotation from the Bible.

The first 'childbirth' recorded in the Bible was that of Eve herself, for she was born of Adam's rib. And how did that childbirth come about ? It is written in Genesis 2:21, 'And the Lord God caused a deep sleep to fall upon Adam, and he slept: and he took one of his ribs, and closed up the flesh instead thereof.'

In short, said Simpson, God had used anesthesia.

Actually, I am not impressed with the counterquotation. Eve was formed while Adam was still in the Garden and before he had eaten of the fruit and, therefore, before sin had entered the world. It was only after the fruit had been eaten that sin and pain entered the world. Simpson's argument was, therefore, worthless.

It was just as well it was, too, for to defeat superstition by superstition is useless. What really defeated the forces of mythology in this case was a revolt by women. They insisted on anesthesia and refused to go along with a curse that applied to them but not to the divines who revered it. Queen Victoria herself accepted anesthesia at her next accouchement and that settled *that*.

Then came 1859 and Charles Robert Darwin's *Origin of Species*. This time the forces of superstition rallied for the greatest battle of all and the preponderance of power seemed on their side. The field of battle was ideally suited to superstition and now, surely, science would be defeated.

The target under attack was the theory of evolution by natural selection, a theory that struck at the very heart and core of human vanity.

It was not a verifiable statement to the effect that a piece of metal would protect man against lightning or that a bit of vapor would protect him against pain that was being considered this time. It was, rather, a thoroughly abstract statement that was dependent upon subtle and hard-to-understand evidence that made it seem that man was an animal much like other animals and had arisen from ancestors that were apelike in nature.

Men might fight on the side of science and against superstition in order to be protected from lightning and from pain for they had much to gain in doing so. Surely they would not do so merely in order to be told they were apes, when the opposition told them they were made 'in the image of God.'

The prominent Conservative Member of Parliament, Benjamin Disraeli (later to be prime minister), expressed the matter so succinctly in 1864 as to add a phrase to the English language. He said, 'Is man an ape or an angel? Now I am on the side of the angels.'

Who would not be?

For once, it seemed, science would have to lose, for the public simply was not on its side.

Yet there were not wanting men to face down the angry multitude and one of them was Thomas Henry Huxley, a largely self-educated English biologist. He had been against evolution to start with but after reading *Origin of Species*, he cried out, 'Now why didn't *I* think of that?' and took to the lecture platform as 'Darwin's Bulldog.'

In 1860, at a meeting of the British Association for the Advancement of Science, at Oxford, the Bishop of Oxford undertook to 'smash Darwin' in public debate. He was Samuel Wilberforce, an accomplished orator, with so unctuous a voice that he was universally known as 'Soapy Sam.'

Wilberforce rose to speak and for half an hour he held an overflow crowd of seven hundred in delighted thrall, while Huxley somberly waited his turn. And as the Bishop approached the end of his speech, he turned toward Huxley and, muting his organ tones to sugar-sweet mockery, begged leave to ask his honorable opponent whether it was through his grandmother or his grandfather that he claimed descent from an ape.

At that, Huxley muttered, 'The Lord has delivered him into my hands.' He rose, faced the audience, and gravely and patiently waited for the laughter to die down.

He then said: 'If then, the question is put to me, would I rather have a miserable ape for a grandfather, or a man highly endowed by nature and possessing great means and influence, and yet who employs those faculties and that influence for the mere purpose of introducing ridicule into a grave scientific discussion – I unhesitatingly affirm my preference for the ape.'

Few debates have ever resulted in so devastating a biter-bit smash and the last offensive against science by superstition was condemned to defeat from that moment.

Huxley had made it clear that it was science now that spoke with the thunders of Sinai, and it was the older orthodoxy that, in the fashion of Wilberforce's unfortunate remark, was capering about the golden calf of man-made myth.

The fight did not end, to be sure. Disraeli was still to make his own unctuous remark, and pulpits were to thunder for decades. I am still, even in this very year in which we now live, frequently made a target by sincere members of the Jehovah's Witnesses sect, who send me publication after publication designed to disprove the theory of evolution.

But the real battle is over. There may be skulking skirmishes in the backwoods and it may even be incumbent upon the astronauts of Apollo 8 to stumble their way haltingly through the first few verses of Genesis 1 as they circle the Moon (in an absolute masterpiece of incongruity), but no man of stature from outside science arises to denounce science.

When science threatens mankind with danger, as in the case of the atom bomb, or bacteriological warfare, or environmental pollution; or when it merely wastes effort and resources as (a few maintain) in the case of the space program, the warnings and criticisms are mounted from within science.

Science is the secular religion of today and scientists are, in a very literal sense, the new priesthood. And it all began when Ben Franklin flew his kite in a thunderstorm in the crucial year of 1752.

15

The Sin of the Scientist

Recently an article appeared in a science fiction fan magazine entitled 'Asimov and Religion.' Written by a young enthusiast named M. B. Tepper, it analyzed certain of my stories in an attempt to show the vein of sincere religious thought that ran through them.

I was astonished, really, for I had not noticed that myself. Actually, I don't practice the rituals of any organized religion and I do not attend any houses of worship. I am a strict rationalist and tend to dismiss anything that does not fall within the scope of reason, as I judge that scope to be.

However (and this is what may mislead people), I am *interested* in religion intellectually, as I am interested in almost everything, and I know a fair amount about the religions of the Western world. What's more, I have no objection to saying things in a theological fashion if that offers economy and sharpness in presenting a thought.

Thus, I was much impressed by J. Robert Oppenheimer's remark, in connection with the invention of the nuclear bomb, that there 'physicists have known sin.' I consider it a sharp and dramatic way of saying a great deal in four words.

I would therefore like to take that phrase as my text (to continue the theological cast of thought) and expound upon it.

In particular, I want to offer some thought on (1) how 'sin' might be defined in connection with science; (2) when it was that science may have committed its first sin; and (3) whether any particular scientist or scientists may be picked out as the original sinners.

First, the definition of sin, and to begin with we had better look at its proper meaning. It is a theological term and represents a transgression of a divine command; a disobedience to the moral law. In this way, sin is infinitely worse than the mere transgression of a man-made law or the mere violation of common sense. Sin is much worse than those actions which we can call 'crimes,' 'blunders,' 'misjudgments' and so on.

If we want to deprive sin of its theological implications and apply it to science, let us at least keep the strength of the word, and decide to use it for *major* wrongdoing, something for which lesser words such as 'crime' would be inadequate.

Sin, in connection with science, then, might be defined as representing the worst thing a scientist, in his role as scientist and not as mere man, can do.

We might say, for instance, that since it is the job of a scientist to investigate nature and to increase, if he can, the sum of human knowledge and understanding of the Universe (including man, as part of the Universe) then the worst thing a scientist can possibly do is to subvert this aim of science. His maximum evil would be to deliberately misinterpret nature in order, out of sheer malice, to decrease the world's store of knowledge.

But this must be eliminated for there is no way, without divine insight, to be sure anyone is doing this. It is too subjective an evil to be judged. Thus, I can't point to a single case of such behavior in the history of science!

There have been scientists in plenty, to be sure, who have held back the advance of science. Abraham G. Werner held back the advance of geology through his mistaken adherence to the tenets of neptunism despite tons of evidence against it. Jöns J. Berzelius held back the advance of organic chemistry by his stubborn insistence on an inadequate theory of molecular structure. George B. Airy delayed the discovery of Neptune through something approaching negligence.

In all these cases, and many more, however, the scientist in question sincerely thought he was acting for the best. All were devoted to the cause of advancing knowledge and all were quite certain that their every action was meant to uphold that cause. We lose patience with them now only through the glorious clarity of hindsight.

Through hindsight we can accuse scientists of the past of blunders caused by stubbornness, by self-love, by the crotchets of age, and by lack of imagination. All this is bad enough, heaven knows, but they are only blunders and which of us is free of that? Which of us would dare submit himself to the judgment of next century's hindsight in the serene confidence of having all his decisions found wise and correct?

No, we cannot consider sin as involving something within the soul of a scientist. Being men and not God we cannot judge the soul.

Let us then seek a method of judging sin by some objective method that does not require that we see within a man. If we eliminate the possibility of doing deliberate harm to the cause of science itself; then the next step is to consider the possibility of doing harm to mankind.

In the course of advancing knowledge, might it not be possible for a scientist to wreak evil on man? And isn't it much easier to decide whether something will harm man, than whether it will harm something as subtle and abstract as 'science.'

I think so. And as for advancing knowledge harming man, it is almost impossible for it not to do so in some way. Who would quarrel with the decision that the discovery of means of controlling fire was a great thing for humanity? And yet it gave us arson. The stone ax, the spear, the bow and arrow could all be used to help kill game; and it may be that all were originally devised with that in mind. Unless we're vegetarians (and I am not) we can scarcely object to inventions which make meat more available.

And yet the stone ax, the spear, the bow and arrow can be, and, of course, have been, perverted to the killing of human beings.

Gunpowder, too, is not primarily a killer. It was first invented, it would seem, by the Chinese in the Middle Ages and was used for pyrotechnic purposes only. (We still use it for that on Independence Day in the United States, Guy Fawkes day in Great Britain, and general celebrations everywhere.) But in Western Europe in the course of the fourteenth century it was put to use in cannons.

A scientist cannot be responsible, then, for the fact that his advance in knowledge can be perverted to harm to mankind (though of course he ought to do his humble best to prevent it). As long as there is good to be derived from his contribution, he has a right to hope that it is that good which will be derived from it.

To be sure, a scientist might weigh the possible harm against the possible good and decide against his own discovery. Thus, Ascanio Sobrero, an Italian chemist, first prepared nitroglycerine in 1847, was horrified at its explosive properties, and refused to do further work on it out of humanitarian considerations. Alfred Nobel, by mixing nitroglycerine with diatomaceous earth and preparing dynamite, gave mankind something which, for all its destructive potentialities, was extremely useful in all sorts of peaceful and valuable construction work.

In weighing good and bad, there may be misjudgments and blunders, perhaps even crime – but as long as there is the possibility of good, I won't use that ultimately strong word 'sin.'

For a scientist to commit a sin, I would have him devise something or uncover knowledge which can *only* do harm and *cannot* do good. What's more, we would have to be reasonably certain that he *knew* it could do only harm and not good, and that he advanced the knowledge or the device in *order* to do harm.

This makes things pretty extreme and in the entire history of mankind, prior to modern times, I can think of only one case that might possibly qualify.

In the seventh century, an alchemist named Kallinikos (or Callinicus in the Latin spelling) fled from either Syria or Egypt (we don't know which) ahead of the conquering Arab armies and managed to make it to Constantinople.

The Byzantine Empire, of which Constantinople was the capital, was sinking under the hammer blows of the Arabs, who were fired by their new religion of Islam, and in 673, Constantinople itself was besieged.

Constantinople might conceivably have been taken despite the strength of its walls, for the Arabs were strong by both land and sea. The city was having difficulty feeding itself and morale was low.

But then Callinicus came up with the most remarkable 'secret weapon' in the history of military warfare. It was a mixture whose composition is unknown to this day (how's that for a secret!) but which seems to have included naphtha and

189

potassium nitrate so that it would burn with a hot quick flame. Quicklime (calcium oxide) was also added. It reacted with water to yield intense heat so that the mixture kept burning even while floating on water.

This mixture, called 'Greek fire,' was launched upon the water in the direction of the Arabic fleet. It set fire to the wooden ships, but what was worse, the spectacle of fire burning on water horrified the Arabs and destroyed *their* morale. The naval blockade was broken and Constantinople was saved.

Greek fire was purely destructive and I don't know of how it could be turned to constructive use. Callinicus, in inventing it, knew this and devised it for destruction, pure and simple. Yet Callinicus, in doing so, conceived himself to be saving Christianity, and I am sure he thought that harm to Arabs did not count, and that only Christians really counted as humans.

Ever since it has been one form of patriotism or another that has tempted the scientist to sin. 'Yes,' he would think, 'we're doing only harm, but we are harming Them in order to save Us, and the good to Us far outweighs the harm to Them.' Clearly, though, we cannot allow this as a legitimate argument, for if both sides argue in this fashion, we *all* die.

To sharpen matters still further, let's take the matter of sin a step beyond.

As long as an individual scientist does something which blackens his own name only, the deed is not yet as bad as it might be. But what if what he does not only destroy his reputation but stains the very concept of science itself and blackens every scientist who exists? Now *that* would be sin.

But if so, scientific sin could not truly exist until there was such a thing as a concept of science in the aggregate; of science as a system of thought that transcended the collection of scientists who served that system.

This concept did not really exist in very ancient times. To be sure, individual men of ingenuity have advanced man's control over the environment since earliest times. (Someone invented a scheme for the extraction of copper from its ores, for instance.) However, the diffusion of knowledge was so slow that the discovery was soon divorced from the discoverer, and the very consciousness that there had been a discoverer at all was lost.

Early societies therefore attributed skills and inventions to gods. They were gifts. Thus, the secret of fire was not the first child of man's brain, the proud discovery that made him man – it was simply the gift of Prometheus.

The origin of what we might recognize as secular science was with the ancient Greeks in the sixth century B.C. The Ionian philosophers, beginning with Thales, were the first to investigate nature without seeking explanations in the whims of the supernatural; the first to advance the concept of the existence of inexorable laws of nature.

But Greek science dwindled and faded out in the last centuries of the Roman Empire and there followed another period dominated by theology. Even when science began to revive in late medieval and early modern times, there was a theological cast about it, so that scientists were considered evil in terms of literal sin.

Roger Bacon was considered a magician in his own times and shuddering legend was composed of him after his death. The well-known Faust legend was based on the activities of a real alchemist really named Faust.

Their deeds were not the sins of scientists but the sins of sorcerers. Their wickedness consisted not in doing harm to man but in making use of demonic power. They had improper knowledge of the sort man was not supposed to have.

This old fear of the scientist as sorcerer lingered long. As late as the early 1930s, many a science fiction story intoned the grave admonishment, 'There are some things it is not given to man to know.'

Maybe so, but the implication was that this knowledge was not *given*; it had been *forbidden*. It was the fruit of the tree of knowledge all over again. And as long as the scientist sinned in this sense, his sin was one with that of all mankind; for all men similarly sinned.

I maintained in the previous chapter that mankind generally (at least in the Western world) first became aware of science and scientists in a truly secular sense, and even as something in

191

opposition to religion, after 1752, thanks to Franklin's lightning rod. It was only after 1752 that it would make sense to think of scientific sin in non-theological terms and imagine some deed that would blacken science out of purely scientific perversion.

In that case, we must come to the conclusion that scientific sin could only properly be spoken of as potentially existing after 1752. We can throw out the one doubtful case of Greek fire then and ask ourselves the specific question: Was there scientific sin after 1752?

To me the answer is, quite clearly: Yes!

For a long period after 1752, throughout the nineteenth century indeed, science was generally considered the hope of humanity. Oh, there were people who thought this particular scientific advance or that was wicked, and who objected to anesthetics, for instance, or to the theory of evolution, or, for that matter, to the Industrial Revolution – but science in the abstract remained good.

How different it is today! There is a strong and growing element among the population which not only finds scientists suspect, but is finding evil in science in the abstract.

It is the whole concept of science which (to many) seems to have made the world a horror. The advance of medicine has given us a dangerous population growth; the advance of technology has given us a growing pollution danger; a group of ivory-tower, head-in-the-clouds physicists have given us the nuclear bomb; and so on and so on and so on.

But at exactly which point in time did the disillusionment with the 'goodness' of science come? When did it start?

Could it have come at the time when some scientist or scientists demonstrated the evil in science beyond any doubt; showed mankind a vision of evil so intense that not only the scientist himself but all of science was darkened past the point where it could be washed clean again?

When was the sin of the scientist committed, then, and who was the scientist?

The easy answer is the nuclear bomb. It was to that which Oppenheimer referred in his remark on sin.

192

But I say no. The nuclear bomb is a terrible thing that has contributed immeasurably to the insecurity of mankind and to his growing distrust of science, but the nuclear bomb is by no means pure evil.

To develop the nuclear bomb, physicists had to extend, vastly, their knowledge of nuclear physics generally. That has led to cheap radioisotopes that have contributed to research in science and industry in a hundred fruitful directions; to nuclear power stations that may be of tremendous use to mankind, and so on. Even the bombs themselves can be used for useful and constructive purposes (as motive power for spaceships, for one thing). And missiles, which might have hydrogen bombs attached, might have spaceships attached instead.

Besides, even if you argue that the development of the nuclear bomb *was* sin, I still reply that it wasn't the first sin. The mistrust of science itself antedates the nuclear bomb. That bomb intensified the mistrust but did not originate it.

I find a certain significance in the fact that the play *R.U.R.* by Karel Čapek was first produced in 1921.

It brought the Frankenstein motif up to date. The original *Frankenstein*, published a century earlier, in 1818, was the last thrust of theological, rather than scientific, sin. In its Faustian plot, a scientist probed forbidden knowledge and offended God rather than man. The monster who in the end killed Frankenstein could easily be understood as the instrument of God's vengeance.

In *R.U.R.*, however, the theological has vanished. Robots are created out of purely scientific motivation with no aura of 'forbiddenness.' They are tools intended to advance man's good the way the railroad and telegraph did; but they got out of hand and in the end the human race was destroyed.

Science could *get out of hand!*

The play was an international success (and gave the word 'robot' to the world and to science fiction) so its thesis of science out of hand must have touched a responsive chord in mankind.

Why should men be so ready, in 1921, to think that science could get out of hand and do total evil to the human race, when only a few years before, science was still the 'Mr. Clean' who would produce a Utopia if allowed to work?

What happened shortly before 1921? World War I happened shortly before 1921.

World War II was a greater and deadlier war than World War I; but World War I was incomparably more stupid in its details.

Men have made colossal misjudgments in a moment of error and may make more to come. Some day, someone will push the wrong button, perhaps, in a moment of panic or lack of understanding, and destroy the world; but never has constant, steady stupidity held sway for weeks, months and years as among the military leaders of World War I. For *persistent* stupidity, they will never be approached.

A million men and more died at Verdun. Sixty thousand British soldiers died *in a single day* on the Somme while generals thought they could build a bridge of mangled flesh across the trenches.

Everything about the carnage was horrible, but was there anything which managed to make itself felt above that sickening spectacle of mutual suicide? Was it the new explosives used in unprecedented quantities; the machine guns, the tanks? They were only minor developments of old devices. Was it the airplane, first used in battle, in this war? Not at all! The airplane was actually admired, for it was in itself beautiful, and it clearly had enormous peacetime potential.

No, no! If you want the supreme horror of the war, here it is:

On April 22, 1915, at Ypres, two greenish-yellow clouds of gas rolled toward the Allied line at a point held by Canadian divisions.

It was poison gas; chlorine. When the clouds covered the Allied line, that line caved in. The soldiers fled; they had to; and a five-mile opening appeared.

No gap like that had been seen anywhere before on the Western Front, but the Germans muffed their opportunity. For one thing, they hadn't really believed it would work (even though they had earlier experimented with gas in a smaller way against the Russians), and were caught flat-footed. For

194

another, they hesitated to advance until the cloud had quite dissipated.

The Canadians were able to rally, and after the clouds drifted away, their line re-formed. By the time of the next gas attack, all were prepared and the gas mask was in use.

That was *the* horror of World War I, for before the war was over poison gases far more horrible than the relatively innocuous chlorine were put into use by both sides.

So grisly was the threat of poison gas, so insidious its onset, so helpless an unprepared group of victims and, what's more, so devastatingly atrocious did it seem to make war upon *breathing* – that common, constant need of all men – that after World War I gas warfare was outlawed.

In all of World War II, poison gas was not used no matter what the provocation, and in wars since, even the use of tear gas arouses violent opposition. Military men argue endlessly that poison gas is really humane; that it frequently incapacitates without killing or permanent harm; that it does not maim horribly the way shells and bullets do. People nevertheless will not brook interference with breathing. Shells and bullets might miss; one might hide from them. But how escape or avoid the creeping approach of gas?

And what, after all, is the other side of poison gas? It has only *one* use; to harm, incapacitate and kill. It has *no* other use. When World War I was over and the Allies found themselves left with many tons of poison gas, to what peaceful use could those tons be converted? To none. The poison gas had to be buried at sea or disposed of clumsily in some other fashion. Was even theoretical knowledge gained? No!

Poison gas warfare was developed knowingly by a scientist with only destruction in mind. The only excuse for it was patriotism, and is that enough of an excuse?

There is a story that during the Crimean War of 1853–56, the British government asked Michael Faraday, the greatest living scientist of the day, two questions: (1) Was it possible to develop poison gas in quantities sufficient to use on the battlefield? And (2) would Faraday head a project to accomplish the task?

Faraday said 'Yes' to the first and an emphatic 'No' to the

second. He did not consider patriotism excuse enough. During World War I, Ernest Rutherford of Great Britain refused to involve himself in war work, maintaining that his research was more important.

In the name of German patriotism, however, poison gas warfare was introduced in World War I, and it was the product of science. No one could miss that. Poison gas was invented by the clever chemists of the German Empire. And the gas poisoned not only thousands of men, but the very name of science. For the first time, millions became aware that science could be perverted to monstrous evil, and science has never been the same again.

Poison gas was the sin of the scientist.

And can we name the sinner?

Yes, we can. He was Fritz Haber, an earnest German patriot of the most narrow type, who considered nothing bad if it brought good (according to his lights) to the Fatherland. (Alas, this way of thinking is held by too many people of all nations and is not confined to Germany.)

Haber had developed the 'Haber process' which produced ammonia out of the nitrogen of the air. The ammonia could be used to manufacture explosives. Without that process, Germany would have run out of ammunition by 1916, thanks to the British blockade. With that process, she ran out of food, men and morale, but never out of ammunition. This, however, will scarcely qualify as a scientific sin, since the Haber process can be used to prepare useful explosives and fertilizers.

During the war, however, Haber labored unceasingly to develop methods of producing poison gas in quantity and supervised that first chlorine attack.

His reward for his unspotted devotion to his nation was a most ironic one. In 1933, Hitler came to power and, as it happened, Haber was Jewish. He had to leave the country and died in sad exile within the year.

That he got out of Germany safely was in part due to the labors of Rutherford, who moved mountains to rescue as many German scientists as he could from the heavy hand of the Nazi psychopaths. Rutherford personally greeted those who reached

England, shaking hands with them in the fraternal comradeship of science.

He would not, however, shake hands with Haber. That would, in his view, have been going too far, for Haber, by his work on poison gas, had put himself beyond Rutherford's pale.

I can only hope that Rutherford was not reacting out of offended national patriotism, but out of the horror of a scientist who recognized scientific sin when he saw it.

Even today, we can still recognize the difference. The men who developed the nuclear bombs and missile technology are not in disgrace. Some of them have suffered agonies of conscience but they know, and we all know, that their work can be turned to great good, if only all of us display wisdom enough. Even Edward Teller, in so far as his work may result in useful fusion power some day, may be forgiven by some his fatherhood of the H-bomb.

But what about the anonymous, hidden people, who in various nations work on nerve gas and on disease germs? To whom are they heroes?

To what constructive use can nerve gas in ton-lot quantities be put? To what constructive use can plague bacilli in endless rows of flasks be put?

The sin of the scientist is multiplied endlessly in these people and for their sake – to make matters theological once again – all mankind may yet be cursed.

Note:

After this chapter appeared in its original form as a magazine article, I received a letter from an army officer working on biological warfare, who resented the last three paragraphs in particular. He named some useful applications of research in poison gas; such as the use of chlorine in purifying drinking water.

Such applications can be discovered and made use of, however, without preparing the poison gases in tank-car quantities. Of what use are the poison gases themselves in such huge amounts? (Chlorine is quite mild compared to post-1915 poison gases and wouldn't be used in wars – and its peaceful applications were based on earlier knowledge, not gained from poison-gas research.)

What's more, *after* I wrote this chapter, President – as he then was – Richard M. Nixon (by no means noted for flaming liberalism) announced that the United States would abandon biological warfare and would never under any circumstances make use of it. And to force Mr. Nixon to this decision, biological warfare must be horrible indeed and must offer *no* counteradvantages of the slightest.

So I stand pat on the contents of this chapter and wish only I had made it stronger – and that I could somehow convince myself that Mr. Nixon had come across it, directly or indirectly, and that it had helped him come to his decision.

16

The Power of Progression

I live, when I am writing (which is most of the time), in the attic of a suburban middle-class house that is rather on the modest side, but is reasonably comfortable.

I am solvent, always have been, and with reasonable luck, always will be, for I am generously paid for doing what I most want to do in the world. My scale of living is not lavish, for there's nothing much I want out of life beyond a working electric typewriter and a steady stream of blank paper; but what I want, I have, or can get.

I have no boss and no employees, so I am my own master in both directions. My editors are (and always have been) so considerate of my feelings as never to give me a cross word. I am in no trouble with the authorities and (again with reasonable luck) hope never to be.

In short, I live, immersed in my work and in my content, in the richest nation on Earth, in the period of that nation's maximum power.

What a pity, then, that it is all illusion and that I cannot blind myself to the truth. My island of comfort is but a quiet bubble in a torrent that is heaving its way downhill to utter catastrophe. I see nothing to stand in its way and can only watch in helpless horror.

The matter can be expressed in a single word: Population.

There are many who moan about the 'population explosion' but they are rarely specific and their worry is easily shrugged away by the comfortable and indifferent. Population has always been expanding, it would seem, and the standard of living has gone up with it, hasn't it?

After all, more hands and more minds mean more cooperation and more inventiveness, and therefore more progress. A million men can do more than a hundred men can, and their added abilities more than make up for the added difficulties introduced by the interactions of a million rather than a hundred.

And the proof rests in the results. The population of the Earth in 1969 is estimated to be 3,500,000,000 which is far higher than it has ever been in history. Yet the overall standard of living on Earth in 1969 is also far higher than it has ever been in history. This is not to say there aren't hundreds of millions who are constantly hungry; hundreds of millions who are downtrodden, frightened, and enslaved – but in the past it has always been even worse.

Well, then, what are we worried about? Why may we not expect that population and living standard will continue to rise, hand in hand?

That sort of outlook reminds me of the tale of the man who fell off the Empire State Building. As he passed the tenth story, he was heard to mutter, 'Well, I've fallen ninety stories and I'm all right so far.'

Suppose we look at the history of the Earth's population, gathering the best estimates we can find.

Ecologists feel that the preagricultural food supply – obtainable by hunting, fishing, collecting wild fruit and nuts, and so on – could not support a world population of more than twenty million; and in all likelihood the actual population during the Stone Age was never more than a third or half of this at most.

This means that as late as 6000 B.C., the world population could not have numbered more than six to ten million people – roughly the population of New York, Shanghai, or Tokyo today. (When America was discovered, the food-gathering Indians, occupying what is now the United States, probably numbered not much more than 250,000, which is like imagining the population of Dayton, Ohio, spread out over the nation.)

The first big jump in world population came with the introduction of agriculture, when the river civilizations along the Nile, the Tigris–Euphrates and the Indus began, by dint of irrigation, to grow food in quantity rather than to gather it. This made possible the establishment of a much denser population than had hitherto been able to exist in those areas.

The increase of population, thereafter, paralleled the opening of new lands to agriculture. By the beginning of the Bronze Age, the world population may have been twenty-five million; by the beginning of the Iron Age, seventy million.

At the time of the start of the Christian era, world population may have been about 150 million, with one third concentrated in the Roman Empire, another third in the Chinese Empire, and the rest scattered over the rest of the world.

The fall of the Roman Empire meant a local decline of population but the worst of the effects was concentrated in Western Europe and it is doubtful if the world population went down much, if at all. Furthermore, by the year 1000, the invention of the horseshoe, the horse collar and the mold-board plow had made the horse an efficient farm animal so that the cold, damp forest land of northwestern Europe could be cleared and turned to agriculture. By 1600, the world population stood at five hundred million.

European explorers opened up 18,000,000 square miles of new land in the Americas and elsewhere and the Industrial Revolution mechanized agriculture, so that the necessary proportion of farmers to non-farmers began to drop. Agriculture could support more and more people per acre of farm land. By 1800, world population was 900 million; by 1900, it was 1600 million; by 1950, it was 2500 million; and by 1969 it is, as aforesaid, 3500 million.

Looking at these figures, let's consider the length of time it takes to double the Earth population.

Up through A.D. 100, the Earth's population doubled, on the average, every 1400 years. This is an extremely slow rate of doubling when you consider that if every married couple has four children and then dies, the Earth's population would double in a single generation of, say, thirty-three years. Can it be that our prehistoric and ancient ancestors didn't know how to go about having children?

Of course not. They had children with all the facility we display today. The trouble is that most of the children died before their fifth birthday. Growing to maturity was a

201

comparative rarity and even those who made it were lucky if they lived the aforesaid thirty-three years. Life was hard and bitter then, and death was always present.

The inexorable shortness of life is clearly recorded in world literature, but times have changed and we forget and misinterpret.

In the *Iliad*, Homer speaks of Nestor who 'outlived two whole generations of his subjects, and was ruling over a third.' Naturally, we think of him as an ancient, ancient man – but he wasn't. He was probably about sixty; that would have been long enough to bury almost every father and son in his kingdom and to be ruling over grandsons.

Most early societies were ruled by 'elders' of one sort or another. The Romans had their 'Senate,' which is simply from a Latin word meaning 'old' so that a Senator is a Latinized elder. The feeling now, therefore, is that these societies were run by senile (same root as senator) graybeards.

Nonsense! In an early society, anyone who made it past thirty-five was an 'elder.' If you want some interesting corroboration of that, just remember that membership in our own club of ruling elders, the United States Senate, requires a minimum age of thirty. To the founding fathers in 1787, this seemed quite old enough for the purpose. If we were starting from scratch today, I'll bet we would have set the minimum at forty, at least.

Even in Shakespeare's time, the notions of old age were not like ours. *Richard II* begins with the wonderful line: 'Old John of Gaunt, time-honored Lancaster,' so that old Gaunt is invariably presented in any production of the play as a man of about 150, who can just manage to hobble across the stage. Actually, at the time the play opened, good old time-honored Lancaster was fifty-eight years old.

You may think that Shakespeare didn't happen to know that. Well, then, in *King Lear*, the Duke of Kent describes himself at one point by saying, 'I have years on my back forty-eight,' and then later on in the play he is referred to as an 'ancient ruffian.'

We can see, then, why the first divine command to mankind which is recorded in the Bible is 'Be fruitful, and multiply, and replenish the earth . . .' (Genesis 1:28).

If, under ancient conditions, man was not fruitful, he would not multiply. Only by having as many children as possible could he rely on a few surviving long enough to have children of their own.

But times have changed! The earth is replenished and it is no longer necessary to be endlessly fruitful in order to have a few survive. Those who take these words of the Bible, applicable to one set of conditions, and who insist on applying them literally to a completely altered set of conditions are doing mankind an enormous disservice. If I spoke in theological terms, I would say they were doing the Devil's work.

As conditions improved and as the death rate fell somewhat while life expectancy lengthened, the time required to double the Earth's population grew shorter. Here are my estimates for the 'doubling time' at various stages in history:

up to A.D. 100	1400 years
100–A.D. 1600	900 years
1600–A.D. 1800	250 years
1800–A.D. 1900	90 years
1900–A.D. 1950	75 years
1950–A.D. 1969	47 years

You see, then, that it is not merely that the population is increasing that is the worst of it; it is that the *rate* at which population is increasing is itself increasing. That is what makes the situation explosive. And the situation is worse in those areas where it can least afford to be bad. In the Philippine Islands, the current rate of increase implies a doubling time of only twenty-two years.

This decrease in doubling time has been brought about by an unbalanced decrease in the death rate. Birth rates have gone down, too, but not nearly enough to compensate and they have gone down least in the 'underdeveloped' portions of the Earth.

What can we do now?

In order to make some decision, let's get one thing clear. The situation cannot be allowed to continue. I don't mean that the doubling time must not be allowed to continue decreasing. It's worse than that. Doubling time must not even be allowed to stay where it is.

Oh, there are optimists (and in this connection I find it hard to refer to them by that word; I prefer to think of them as idiots) who think that if only we end wars, establish world tranquillity and advance science we can absorb population increase. We need only farm scientifically, make intelligent use of fertilizers, put the ocean to efficient use as a source of food and fresh water and minerals, develop fusion power, harness the Sun— Then we can easily support a *much* larger population than now exists. I have seen statements to the effect that the Earth, Utopically run, could support fifty billion human beings in comfort.

But then what? What's to prevent the population from increasing beyond that? Would not some form of birth control be required then? In other words, the greatest optimist cannot deny the necessity of birth control eventually; he merely says, 'Not yet!'

Is it possible that such an optimist has a dim idea that the time when the Earth's population will reach fifty billion (or whatever generous limit he sets) is so far off that no one need worry now? Or, worse yet, does he have the idea that by the time fifty billion is reached, further scientific advance will make it possible to support still higher numbers and so on into the indefinite future?

If that is so, then the optimist hasn't the faintest idea of the power of a geometric progression. But then, hardly anyone does. Let's see if we can't illustrate that power.

Since the Earth's population is 3.5 billion people and since that population is now doubling at the rate of once every forty-seven years, we can make use of the following equation:

$$(3,500,000,000) \, 2^{x/47} = y \qquad \text{(Equation 1)}$$

This tells us the number of years (x) it will take us to reach a world population of y, supposing that the doubling rate remains absolutely constant. Solving for x in Equation 1, we get:

$$x = 156 \, (\log y - 9.54) \qquad \text{(Equation 2)}$$

Suppose we ask ourselves, now, how long it will be before we reach that population of fifty billion that optimists think Earth can support provided only we establish a Utopia?

Well, if y is set at 50,000,000,000, then $log\ y$ is 10.70 and x is equal to 182 years.

In other words, if the doubling rate continues exactly as now, we will have reached a world population of 50,000,000,000 by A.D. 2151.

The wildest optimism is required if you think that in the space of time in which the American Constitution has thus far existed (six generations) we are going to be able to abolish war and establish the kind of rational Utopia which would make so large a population possible and comfortable.

Even then, we would be much nearer a colossal catastrophe in case anything happened to go wrong with fifty billion people encumbering the Earth than the present 3.5 billion. And what if population continued to increase even past the fifty billion mark? Could we still rely on science to continue to make higher populations possible. How high can populations go in the reasonable future?

Let's move on and see —

The island of Manhattan has an area of 22 square miles and a population of 1,750,000. In the middle of the working day, when people come to Manhattan from adjoining areas, the population jumps to 2,200,000 at least; at which time the population density is 100,000 people per square mile.

Suppose all the Earth were covered with people as thickly as Manhattan is at lunchtime. Suppose the Sahara Desert were covered that thickly, and the Himalayan Mountains, and Greenland and Antarctica and everywhere else. Suppose we threw planks over all the oceans and crowded those planks like a Manhattan lunch hour as well.

The Earth's total surface area is 200,000,000 square miles. If all of it were populated at Manhattan density, the world population would be 20,000,000,000,000, or 20 trillion. How long, now, would it take to reach that figure?

As Equation 2 would tell you, the answer is the astonishingly small one of 585 years. By A.D. 2554, at the present rate of increase, Earth's surface will become one huge Manhattan.

Of course, you may decide not to let me get away with that. I am, after all, a science fiction writer at times, and I know all about space travel. Surely by A.D. 2554 men will be flitting all over the solar system and will therefore be able to populate the planets, which then will be able to absorb some of the population excess from Earth.

Sorry, but that's not good enough. In the next 47 years, we would have to export 3.5 billion people to the Moon and Mars and wherever, just in order to stay where we are now on Earth. Is there anyone here who thinks we can do that in 47 years? Is there anyone here who thinks the Moon and Mars and wherever can be engineered to support 3.5 billion people in the next 47 years even if we could get them there?

In fact, let's go further. There are about 135,000,000,000 stars in the Galaxy. Some of them may have habitable planets in the sense that men could live on them without prohibitive engineering.

Of course, we can't reach such planets, either now or in the foreseeable future, but suppose we could. Suppose we could transfer human beings instantaneously to any planet we wished by a mere snap of the fingers and with no further expenditure of energy than that. And suppose that there was an incredible wealth of habitable planets in the Galaxy; suppose that every single star in the Galaxy had ten such planets. There would then be 1,350,000,000,000 habitable planets in the Galaxy.

Suppose further that the same were true of every other galaxy and that (as some estimate) there are a hundred billion such galaxies in existence. This means there would be 135,000,000,000,000,000,000,000 habitable planets altogether.

Finally, what if we continued snapping fingers and transferring people until every one of these planets was populated to Manhattan density. The total population of the Universe would then be 2,700,000,000,000,000,000,000,000,000,000,000,000 or 2.7 trillion trillion trillion.

How long would it take us to reach such a population, eh? Now that we are talking of trillions of trillions of trillions of people, it might seem that we can wait many millions of years to fill the Universe in this impossible way. If you think so, you still don't understand the power of a geometric progression.

At the present rate of population increase, it will take us only 4200 years to reach a population of 2.7 trillion trillion trillion.

By A.D. 6170, we will have crammed the Universe from end to end with people. Every star in every galaxy will see each of its ten planets carrying a population that will resemble a Manhattan rush hour on every part of its surface.

Do you think I can't get any more extreme? Suppose man's scientific advance managed to turn all the Universe into food and to tap hyperspace for energy. How long would it take for the entire mass of the known Universe to be turned into human flesh and blood? The Sun has a mass of 4.4 million trillion trillion pounds. Estimate the average weight of a human being at 110 pounds and you find that if the Sun were converted into people, it would make up a population of 40,000 trillion trillion.

Multiply that by 135 billion to convert the Galaxy into people; multiply that again by 100 billion to convert all the galaxies into people; multiply that again by 100 to account for the dust and debris that exist in the Universe outside the stars and the total mass of the Universe converted into people makes for a population of 54,000,000,000,000,000,000,000,000,000,-000,000,000,000,000,000,000,000, or 54,000 trillion trillion trillion trillion.

How long will it take our progression to reach that? By now, you should have no high hopes. It will take 6700 years. By the year A.D. 8700, we will have run completely out of Universe and that's all there is to it.

Science, in other words, cannot keep up with population for very long no matter what it does.

It's an absolute certainty that we are not going to multiply at our present rate till we consume the entire Universe or even till we merely fill the surfaces of all the planets. I think you will agree that the extreme of optimism will not carry us past the conversion of Earth itself into one big Manhattan. This means we have as our outside limit, the year A.D. 2554. We have only a little over five and a half centuries left us.

Whatever happens to make the rate of population increase smaller, or abolish it altogether and give us a stable population, *must* happen before A.D. 2554. I don't say 'should' or 'ought to' or 'might.' I very deliberately say 'must.'

But do we really even have that much time? What does a planet-wide Manhattan mean?

The total mass of living objects on Earth is estimated at 20 trillion tons, while the present mass of humanity on Earth is about 200 million tons. This means that humanity makes up 1/100,000 of the total mass of life on Earth. That's pretty good for a single species.

All of life is supported by plant photosynthesis (with some insignificant bacterial exceptions). Animals can only survive by raiding the chemical energy (food) built up by plants out of solar energy. Even those animals who eat animals only live because the eaten animals ate plants – or if they ate animals too, then those animals ate plants. However far the chain extends, it comes to plants in the end.

It is estimated that the total mass of an eater in a food chain must be only one tenth the total mass of the eaten, if both are to survive at a stable population level. This means that all of animal life has a mass of 2 trillion tons and the mass of humanity is 1/10,000 of that.

Since radiation from the Sun is a fixed quantity and the efficiency of photosynthesis is also fixed, only so much animal life can be supported on Earth. Every time the human population increases in mass by one ton, the mass of non-human animal life must decrease by one ton to make room.

How long, then, will it take the human race to increase to the point where its mass is equal to the maximum mass that animal life may have? The answer is 624 years. In other words, by the time all the Earth is Manhattanized, we will have had to kill off just about all of animal life. All the remaining wild life will be gone. All the fish in the sea, all the birds in the sky, all the worms underground; even all our own domestic animals and pets, from horses and cattle to cats, dogs and parakeets, will have to go, sacrificed at the altar of human procreation.

(Think of that, you conservationists, and remind yourselves frequently that while human population increases, animal life *must* dwindle and not all your piety, wit or tears can do anything about it. If you want to fight the good fight for conservation, fight the better fight for population control.)

What's more, killing off animals is only part of it. All plant life would have to be converted into food plants, with as little non-food portion as possible. In the day when the Earth becomes one large Manhattan – one large planet-girdling office building – the only living things on Earth other than human beings will be those little cells in the algae tanks all over the roof of that building.

Theoretically, we could learn to utilize solar energy and convert it into synthetic food without the intervention of plants, but do you think we can work this out at a level necessary to support a population of twenty trillions within the next five or six centuries? I don't.

Nor is it only a matter of food. What about resources? Already, with a population of 3.5 billion and the present level of technology, we are eroding our soil, spreading our minerals thin, destroying our forests, and consuming irreplaceable coal and oil at a fearful rate. Remember, that as the population increases, the level of technology and therefore the consumption of resources increase even faster.

And what about pollution? Already, with a population of 3.5 billion and the present level of technology, we are poisoning the land, sea and air to a dangerous extent. What will we be doing in a century when the population is 14 billion?

These problems are perhaps not insoluble if we let them grow no worse, but they would even then be soluble only with great difficulty. How will they be solved if the resource expenditure and the waste production grew worse with each year, as they are doing and will continue to do.

Finally, what of human dignity? How decently can we live when crowds of human beings and their tools clog every highway, every street, every building, every piece of land? The human friction that results when space disappears and privacy is destroyed makes itself evident in increasing discontents and hatreds, and this friction will grow phenomenally worse as the population continues to multiply.

No, take it all in all, I don't see how we can dare let mankind increase at its present rate for even a single additional generation. We must reach a population plateau in the early decades of the twenty-first century.

And I'm sure we will, one way or another. If we do nothing but what comes naturally, the population increase will be brought to a halt by an inevitable rise in the death rate through the wars and civil rioting that worsening human friction and desperation will bring; through the epidemics that crowding and technological breakdown will bring; and through the famines that food shortage will bring.

The reasonable alternative is to reduce the birth rate. That, too, will fall, naturally, when crowding and starvation make human procreation less efficient, but do we want to wait for that? If we wait for that, the famines will start in places like India and Indonesia (I predict) by 1980.

Let me summarize as bluntly as possible. There is a race in man's future between a death-rate rise and a birth-rate decline and by 2000, if the latter doesn't win, the former will.

17

My Planet, 'tis of Thee —

I love people, I really do, and yet, in viewing the future, I am forced to be guided by a certain cynicism because so many people, however lovable, seem immune to reason.

Several years ago, for instance, at some gathering, a Jewish woman argued, with considerable emotion, that she could never feel real confidence in the good will of Gentiles because they had stood aside and allowed Nazi Germany to torture and kill Jews by the millions without ever really doing anything about it.

I could appreciate her feeling, being Jewish myself, but didn't share them. To make my point, I asked her quietly, 'What are you doing about Negro civil rights?'

And she answered sharply (and rather as I expected), 'Let's solve our own problems before we take on those of other people.'

But I had not made my point after all for – would you believe it – there turned out to be no way in which I could convince her of the inconsistency in her position.

But we have to take people as they are; complete with their aversion to the rational; and face, in these last decades of the twentieth century, the most crucial problem mankind has ever had to deal with. It is the question of sheer survival; not for this sect or that, this nation or that, this political or economic doctrine or that – but, quite simply, for civilization generally.

And maybe even for mankind generally.

And maybe even for multicellular life generally.

The prime problem is that of increasing population, something I considered forcefully in the previous chapter, and am taking up even more forcefully in this final chapter. Even if, right now as of this minute, the population of Earth levels off, we still face an overwhelming problem. The population of Earth is *already* too high for survival, for we grow in other ways than mere numbers.

You see, it is an article of faith with us that we must live in a 'growing economy,' that we must 'progress' and 'advance'; that we must have it ever better. Arguing against all that is like arguing against kindness and mercy and love, but I have to. Growing, and advancing and having it better have a price tag. To have a still more affluent society inevitably means the utilization of Earth's resources at a still greater rate and, in particular, the consumption of energy at a still greater rate.

As it is, the utilization of irreplaceable resources and the consumption of energy have been increasing faster than the population for many decades, and I have seen it stated that by the time the United States has doubled its population, say by 2020, it will have increased its energy consumption seven-fold.

I strongly suspect that the rate of pollution of the environment is roughly proportional to the rate of energy consumption and that, barring strenuous action to prevent it, a sevenfold increase in the latter means a sevenfold increase in the former. And in only fifty years.

Look at it another way.

As it happens, the United States is currently consuming some-what over half the irreplaceable resources produced on Earth – the metals, fossil fuels and so on – despite the fact that it only has one sixteenth the population of the Earth.

The rest of the world would, of course, like to attain our level of affluence and it is hard to argue that they have no right to do so. But suppose they succeed and all Earth lives on the American living standard. The remaining fifteen sixteenths of the Earth's population would then be using fifteen times as much of Earth's resources as our one sixteenth does, and will produce fifteen times as much pollution.

The rate of coal and oil production, then, of metal and mineral production, of paper, plastics, of automobiles and everything else would have to be something like eight times what it is now to keep all of Earth at American-like affluence even if the population increases no higher than it is right now. And eight times as much pollution.

Can we afford this?

And if population continues to increase at its present rate

and world affluence is *still* expected, then in fifty years, with a sevenfold increase in energy consumption expected, the rate of rifling of Earth's resources must be over *fifty* times what it now is.

It can't be done. The resources simply don't exist for the rifling. The capacity for the absorption of a fiftyfold increase in pollution just isn't there.

The fact of the matter is that we can no longer proceed along the lines that have served us in the past. We can't imagine that we can continue to increase our rate of production just as fast as we can manage, that unlimited growth is possible (let alone desirable) and that good old Earth will give us all we need, however much that is, and take all the dregs we hand back, however much that is.

We are *not* in an infinite world any longer; we are (and have been for some time now) in a terribly *finite* world; and we must either adjust to that or die.

We can measure the finiteness of a world by its interdependence and we can trace the growth of interdependence by a rough estimate of the value of distance at various times in history. Thus, as long as the Earth was essentially infinite, it was possible for particular portions to be far enough apart to ignore each other; and the greater the distance required for such ignoring the closer the approach of finiteness.

Just to take a few examples—

In 1650 B.C., it did not concern the Greeks that the Egyptian Middle Kingdom (five hundred miles away) had fallen to the Hyksos invaders. In 525 B.C., however, the fall of Egypt to Persia clearly heightened the dangers besetting Greece.

In 215 B.C., the deadly duel between Rome and Carthage raised no echo in the hearts of Britons on their tight little isle, one thousand miles away. By A.D. 400, however, the state of Italy with respect to Germanic invaders was of intense interest to Britain, for Alaric's presence in northern Italy cost Britain its Roman army – and its civilization.

By A.D. 1935, most Americans could still live as though it didn't matter what happened in Europe, three thousand miles away, but after less than another generation, they would be told and believe, that what happened in Saigon, ten thousand miles away, was of such vital importance to Keokuk that tens of thousands of Americans must die.

If one wanted to take the trouble, one could extend this thing backward in time and fill in the chinks and work out a graph representing the change in minimum distance of separation required for isolation, with time. The line would not be straight but I think we can all see that it would rise more or less steadily; quite slowly at first; quite rapidly later on.

And now we've run out of distance. The maximum distance of separation of Earth's surface is 12,500 miles and that is not far enough any longer to be safe. Oh, people might want to be isolated and might stubbornly insist they don't care what happens 12,500 miles away or even 500 miles away, but when they say that they are simply closing their eyes at noon and insisting that the Sun has set.

The interdependence is not just political and military, of course, but economic, social, cultural and everything else. Nothing of significance (and hardly anything any longer is of insignificance) that happens anywhere can fail to affect everywhere. Because four rock singers in Liverpool decided a few years ago that it was a drag to go to the barber, I now wear sideburns.

But never mind the trivial. Let's go right to the top and make one thing clear. It is no longer possible to solve the real problems of our planet by working on the assumption that the world is infinite. Whatever the problem is, whether it is overpopulation, dwindling resources, multiplying pollution or intensifying social unrest, no minor part of the planet can solve it without regard for, or attention to, the rest of Earth.

Since no single nation controls an effective majority of Earth's population, area, resources or power, each nation comes under the heading of 'minor part of the planet.' It follows, in my opinion, that no nation can solve its own major problems alone.

Paraguay cannot solve its major problems in isolation, and neither can the United States, and for the same reason. Both Paraguay and the United States deal with too small a part of the problem, and anything either does, can and will be negated if the rest of the planet does not choose to go along.

Suppose the United States did decide to enforce a policy of

strict population control and established a firm population plateau at two hundred million. Can this possibly solve our problems under any conditions, if the rest of the world continued to breed itself upward like rabbits?

One might, for the sake of argument, maintain that with a stable population and our superb technology, we could easily defend ourselves against the rest of the world, especially if the rest of the world sinks into starvation and disaster through overcrowding.

But some of them out there have nuclear weapons that can do us tremendous damage while we're wiping them out; and all of them have resources that we have to collect. We can harden our hearts to the famines they will undergo but can we harden our immunity to the plagues that will sweep them on the way to us?

And what if the rest of the world manages to achieve population stability also and continues to drive hard for greater economic affluence, each nation in a mad scramble for it without regard to the others? Can we tell them that half the resources they control belong to us and that they'd better not use those resources recklessly because we don't intend to give up our affluence for their sake? Our record in Vietnam doesn't fill me with confidence in our power to impose our will on the world.

And if they improve their lot somehow without disimproving ours and begin to pollute at the rate one would expect of affluence, what then?

If we in the United States cleaned up our rivers and unclogged our lakes and de-stinked our atmosphere, what super-Canute is going to say to the air and water of the Earth generally: Cross not our borders for you are unclean?

In short, our problems are now planetary, and our solutions will have to be planetary too.

What this means is that the various nations have to come to some sort of agreement in order to find and implement planetary solutions.

As a barest minimum, the United States and the Soviet Union will have to work together. Not only is war between them unthinkable, but disagreement, short of war, out of national stubbornness and suspicion, that will prevent common action on these problems will only mean a somewhat slower and perhaps, in the long run, more agonizing death than war could bring.

Pride and patriotism aren't going to work on a national basis. It is utterly irrelevant that we think the Russians are a bunch of Commies or that they think we are a bunch of Imperialists. Whether they're right, or we're right, or both of us are right or neither, makes no difference. We would still have to work together, agree on a common policy, and stay agreed on it, or all go under.

Even the United States and the Soviet Union together might not swing it. It might require the wholehearted cooperation of Western Europe, China (Communist China, not Formosa) and India as well, to set up the minimum requirements for planetary solutions. The rest of the world could then follow voluntarily or reluctantly or even under the lash – but follow.

Don't get me wrong. I don't enjoy the thought of the powerful nations of the world getting together to force the rest of the planet into line. I would love to see a democratic world government with a strong, freely elected executive with limited tenure, a representative legislature, respected world courts, and a firm agreement on the part of all peoples to abide by majority decisions. The only trouble is I don't think we have the faintest chance of swinging it in thirty years, and thirty years are all the time we have in my opinion.

In the short run, therefore, we must settle for something less.

The trouble is that our society, our culture, our every way of thinking is built around the assumption of an infinite world. It is the very essence of my lady friend of the introduction and her: 'Let's solve our own problems before we take on those of other people' that she doesn't see that there are no longer such things as my problems and your problems – but only problems.

It is not inconceivable, therefore, that under the dark glimmering of outmoded ways of thought, we will all sink to death rather than cooperate, for we will feel that *those* bad guys may use that cooperation to take advantage of *us* good guys. (And those guys would be saying exactly the same thing with the pronouns reversing referents.)

With this possibility horribly easy to see coming to pass and with an ideal world government almost certainly far in the future, I will be ecstatically happy to see cooperation on any terms. Let every nation desperately pretend it is retaining its sovereignty; let it bellow its resentment and hatred for the others; as long as each nation cooperates (even sneakily and sullenly) in measures that are even half good and manage to keep us drifting along with our nostrils above water until such time as the real world government can be developed.

There is a historical analogy I would like to offer. It is imperfect and inadequate, as all historical analogies are, but here it is anyway.

In 1776, the rebelling British colonies on the North American east-central coast declared themselves 'free and independent states' and were recognized as such by Great Britain in 1783.

Through long habit we think of states as parts of a nation, marked off for administrative purposes and possessing no true independence, but if so we are deluding ourselves. The word 'state' refers to a self-contained political entity, entirely self-governing. The thirteen states, free and independent, were filled with the mutual suspicions you would expect of any neighboring governments. For several years after 1783, the United States were states that were united only in their title. They fought with each other propagandistically, economically and, very nearly, militarily.

The tighter union, under the Constitution (prepared in 1787, adopted in 1789 – with Rhode Island holding out till 1791), was by no means an enthusiastically accepted situation. It was forced upon the mutually hostile states, to their reluctance and chagrin, by the existence of Continental problems that required Continental solutions, failing which, European overlordship would surely have been reimposed.

Nor was the Constitution more than jerry-built at first. It was the states that were 'sovereign,' their acceptance of Union was thought to be voluntary and at different times various states (suffering under a feeling of being abused) played with the notion of withdrawing that acceptance and leaving the Union.

Finally, in 1860 and 1861, such withdrawal was actually

attempted by eleven states of the South and it took four years of bloody warfare to point out the fallacy of their reasoning.

The situation of 1787 is being repeated now on a planetary scale in area and population, and considering the size of the first and the heterogeneity of the second, we might consider a happy ending this time as beyond hope, but I wonder—

In some ways difficulty has lessened. Advances in transportation and communication have made the entire planet a far smaller place in every way than the eastern seaboard of the United States was in 1787. And the heterogeneity of the world's population is less than one might think. Tokyo and Cairo look more like Chicago in 1970, than Charleston looked like Boston in 1787. Of course, there are differences and great ones, but the ruling classes in all nations live more and more as part of the same (largely Americanized) culture.

So planetary union by paste and piano wire, while the pretense of national sovereignty is maintained, can come. But will it last?

How long did the American Constitution serve to hold the American states together? Seventy years.

Then came the bitterest and most stubbornly contested civil war in Western history, and it would not have taken much for it to end the other way, with the Union destroyed.

Fortunately for us, the world had not yet become finite in the nineteenth century. The United States could fight a civil war and survive.

But over a century has passed since then and we can no longer take such risks. Suppose a working planetary government is set up, conceived in expediency and dedicated to the proposition that mankind must somehow survive.

If, then, seventy years later, there is a planetary civil war, do you suppose we will have to see which side wins in order to find out whether any planetary government so conceived and so dedicated can long endure? Of course not; the mere fact that there is civil war at all would settle the matter. The government, civilization, mankind would not endure.

So in the decades that follow the beginnings of foot-dragging cooperation, we must go on to establish greater sympathy among peoples and steadily diminish idiotic national prejudices

218

and do so without a single false step. That is the price of finitude.

It won't be easy, but it's got to be done.

Well, how?

Shall we try another imperfect historical analogy?

It is my feeling that what did most to break down the sharp feeling of statism within the United States was the westward migration. The opening of the Western frontier was the common task of the settled states of the East. The West was not parceled out, area by area, for this state and that, so that each of the older states might increase its pride in its specialness and work up further ground for hatred against the rest. All the West was open to all Americans.

Men from the various states mixed freely and no significant portion of the task of 'the winning of the West' could be attributed to anything smaller than the United States in general.

Sure, regional pride continued and will always continue, but it was defused to the point where members of one state do not feel they have a God-given right to kill members of another state.

And what is the equivalent of the westward migration on a planetary scale?

How about the space effort?

It is getting hard to say this. The entire program has been so nationalized by the traumatic effect of the Soviet Union having been the first nation to place a satellite in orbit that it has become the darling of the conservatives. It has been militarized, it has been draped with the flag, it has been permeated by an aura of *Reader's Digest* and Billy Graham.

So it has become an object of suspicion to the liberals.

To many of the latter it now seems like an opiate, designed to keep the eyes of the American people fixed on the Moon while on Earth the cities decay and people fester. Over and over again they tell us the choice is Moon or Earth, space or cities, rockets or people.

If that were really the choice, I would choose Earth, cities and people myself, but it isn't. The real problem is that almost every nation on Earth spends most of its money and effort on preparing for war (or actually fighting one). The choice is not Moon or Earth at all. The choice is war or Earth, soldiers or cities, missiles or people, and every nation chooses war, soldiers and missiles.

Achieve a world government of any kind, however rickety, and let military expenditures die down and there will at once be enough muscle and brain extant to make possible both space and cities, both Earth and Moon, both people and rockets.

But why bother? What's the good of a man on the Moon?

I have argued in other places about the material good that might come of it; of knowledge to be gained that will increase our understanding of Earth's geology, of the solar system's origin, even of the workings of life. I have spoken of the new technology that might become possible based on the abundance of vacuum, hard radiation and low temperature on the Moon.

I have even argued in favor of the establishment of an ecologically independent colony on the Moon; one that (after its initial start with Earth capital) can continue on its own through the careful utilization of resources available from the Moon's crust – pointing out that such a Moon colony can offer Earth an abstract service far beyond any concrete advantage it might bring.

The reason for that is that the Moon is so obviously a finite world. Its lack of surface air and water means that any colonists on the Moon must cycle their resources with infinite care, for they will be living in a society that will leave almost no margin for error.

Assuming that the colony survives and functions, then it will serve us both as an inspiration and as an example. It will show us how human beings can live with finiteness, and it may even help teach us how.

But forget all that. Let us suppose that a real Moon colony turns out to be impossible; that the material benefits of space turn out to be an illusion; that the knowledge gained by scientists is of use to theoreticians only and remains worthless to the common man. Let us suppose that the space effort, both now and in the future, is simply a vastly expensive boondoggle, a mere climbing of super-Everests.

It would, even so, *still* be infinitely worthwhile.

As it stands, the space effort has grown too expensive for any single nation – even for the United States or the Soviet Union – that insists on spending most of its effort on the sterility of armaments and war.

Once the more powerful nations, at least, are forced by dreadful circumstance into cooperation and the makeshift planetary government starts, the space effort will, almost inevitably in my opinion, become a multinational effort.

And there lies my greatest hope for the survival of mankind.

Why should not all nations find a common ground in the assault on space? The only enemy there is the dark of the unknown and surely that is an enemy all mankind can fight with equal enthusiasm. The exploration of space can call upon us all alike to defeat ignorance, to open new horizons, and it will present mankind with a kind of accomplishment so large as by its very size to shrivel nations to insignificance and leave room in the mind for nothing smaller than planetary man.

Even now there is a glimpse of the supranational attraction of space accomplishment. The successful placing of a man on the Moon was, admittedly, a strictly American achievement, full of American flags stuck into lunar soil, and American Presidents hastening to the Pacific Ocean with their benediction, and American Vice-presidents handing out bits of Moon rock to everyone in Asia – and *still* the accomplishment touched everyone, even the Russians. For it was Homo sapiens that was making footprints on the Moon and it was for all of us without exception that those footprints stood for.

Let us have the space effort become multinational in the future, let there be a planetary flag thrust into the soil of Mars, let men of all parts of the world work on the vast project for the conquest and taming of the solar system, and surely the consciousness of our union-as-a-species will have a chance to grow.

Again regional rivalries will remain, always, but overriding them (just possibly) may be the sense of common accomplishment that will slowly but surely break down the disunion of men and leave them with just enough resigned endurance of one another (in the absence of love, resigned endurance will be enough) to turn a makeshift world government into a real one.

(It may not happen, to be sure, for the exploration of the New World in the sixteenth century did not bring the European nations together but exacerbated their rivalry – but then they never made it a multinational project.)

If it does happen, however, the space effort, whatever its cost, short of planetary bankruptcy, will have been worthwhile even if it brings us nothing else.

Also, if that happens, the twenty-first century will see mankind making the painful transition from the childhood of a pseudo-infinite world of subplanetary societies, through the adolescence of a cooperative national society and into the adulthood of a planetary government ruling over a finite world.

The chances for all of this are, I repeat, not large, for time is short and folly long, the need is great and vision small, the pressing problems enormously complex and the ruling minds dishearteningly mediocre.

But I must hope.

Isaac Asimov, Grand Master of Science Fiction, in Panther Books

The Foundation Trilogy

FOUNDATION	35p ☐
FOUNDATION AND EMPIRE	35p ☐
SECOND FOUNDATION	35p ☐
THE EARLY ASIMOV (Volume 1)	40p ☐
THE EARLY ASIMOV (Volume 2)	40p ☐
THE EARLY ASIMOV (Volume 3)	40p ☐
THE GODS THEMSELVES	40p ☐
THE CURRENTS OF SPACE	35p ☐
THE NAKED SUN	30p ☐
THE STARS LIKE DUST	35p ☐
THE CAVES OF STEEL	40p ☐
THE END OF ETERNITY	35p ☐
EARTH IS ROOM ENOUGH	35p ☐
THE MARTIAN WAY	30p ☐
NIGHTFALL ONE	30p ☐
NIGHTFALL TWO	30p ☐
ASIMOV'S MYSTERIES	35p ☐
I, ROBOT	30p ☐
THE REST OF THE ROBOTS	35p ☐

All these books are available at your local bookshop or newsagent; or can be ordered direct from the publisher. Just tick the titles you want and fill in the form below.

Name..

Address ....

....

Write to Panther Cash Sales, P.O. Box 11, Falmouth, Cornwall TR10 9EN
Please enclose remittance to the value of the cover price plus 15p postage and packing for one book, 5p for each additional copy.
Granada Publishing reserve the right to show new retail prices on covers, which may differ from those previously advertised in the text or elsewhere.